Oxygen Disorder Effects in High-T_c Superconductors

Oxygen Disorder Effects in High-T_c Superconductors

Edited by

J. L. Morán-López
Universidad Autónoma de San Luis Potosí
San Luis Potosí, San Luis Potosí
Mexico

and

Ivan K. Schuller
University of California, San Diego
La Jolla, California

Plenum Press • New York and London

Library of Congress Cataloging-in-Publication Data

Oxygen disorder effects in high-Tc superconductors / edited by J.L.
Morán-López and Ivan K. Schuller.
    p.   cm.
  Includes bibliographical references.
  ISBN-13:978-1-4612-7867-2   e-ISBN-13:978-1-4613-0561-3
  DOI: 10.1007/978-1-4613-0561-3
    1. High temperature superconductors. 2. High temperature
superconductivity. 3. Superconductors--Chemistry. 4. Oxygen.
I. Moran-López, J. L. (José L.), 1950- II. Schuller, Ivan K.
QC611.98.H54O99 1990
537.6'23--dc20   89-49258
                                                          CIP

Proceedings of a conference on Oxygen Disorder Effects in High-Tc
Superconductors, held April 18–21, 1989, in Trieste, Italy

© 1990 Plenum Press, New York
Softcover reprint of the hardcover 1st edition 1990

A Division of Plenum Publishing Corporation
233 Spring Street, New York, N.Y. 10013

PREFACE

The papers in this book represent the proceedings for the International Conference on **Oxygen Disorder Effects in High-Tc Superconductors**, held April 18–21, 1989 at the International Centre for Theoretical Physics, Trieste, Italy.

It was recognized very early in the field of ceramic superconductors that oxygen plays a crucial role as far as the physical properties of these materials are concerned. The preparation requires special heating and cooling cycles which allow proper uptake of oxygen, relationships were found between the oxygen concentration and the superconducting transition temperature in many of the compounds and quite recently it was recognized that many (if not all) of the compounds present oxygen ordering in the intercalating planes.

Moreover, it seems that the presence of superconductivity is strongly correlated with the presence of orthorhombic phases although several groups have also claimed the presence of superconductivity in tetragonal phases. Whether oxygen ordering plays or not a crucial role for the superconductivity remains to be seen. However it is clear that the ordering of oxygens and their thermodynamic properties is an interesting subject on its own right. All these reasons led us to organize a Conference on **Oxygen Disorder Effects in High-Tc Superconductors** in attempt to identify unsolved problems and to have an open discussion of the presently *known* facts.

The general consensus of the meeting was that oxygen is *vital* in the high Tc superconductor materials. From the conference the following points can be marked:

1. The lattice structure as well as the ordering of oxygen in the high Tc superconductors is now well established.
2. The theoretical understanding of the oxygen ordering and of the phases is also good on the basis of phenomenological models.
3. The electronic structure of stoichiometric compounds is very well established from band structure calculations.
4. Tight–binding models for the local density of states of partially ordered systems have been performed and seem appropriate for these systems.
5. Substitution studies show interesting trends but the full interpretation is still unclear.

6. Much experimental data on the magnetic properties of rare earth ions in the high Tc materials as a function of oxygen content were presented. The interpretation of most of them is lacking.

7. Raman scattering experiments for various oxygen concentrations were presented and models for the oxygen vibrations were discussed.

The conference counted with the participation of researchers, theorist and experimentalists involved in issues related to the main team of the conference. Sponsorship was received from the International Centre for Theoretical Physics, the Organization of American States, the Mexican National Council of Science and Technology, to help defray travel and subsistence cost of many of the participants. We thank the participants for all the efforts they have put on the presentations and the manuscripts and to Professor Mario Tosi for his support in the organization of the meeting.

J.L. Morán–López
San Luis Potosí, S.L.P., Mexico

I.K. Schuller
San Diego, California

August, 1989

CONTENTS

MAGNETIC PROPERTIES

CHEMICAL SUBSTITUTIONS

PHONONS

SUMMARY

POWDER NEUTRON DIFFRACTION STUDY OF THE STRUCTURAL PHASE DIAGRAM OF THE $Ba_{1-x}K_xBiO_3$ SYSTEM

Shiyou Pei, J. D. Jorgensen, D. G. Hinks, B. Dabrowski
D. R. Richards and A. W. Mitchell

Materials Science Division
Argonne National Laboratory
Argonne, IL 60439

S. K. Sinha, D. Vaknin, J. M. Newsam and A. J. Jacobson

Exxon Research and Engineering Company
Clinton Township, Route 22 East
Annandale, NJ 08801

In this article we present a brief report of recent work carried out to elucidate the phase diagram of the $Ba_{1-x}K_xBiO_3$ system. A full account of this investigation will be published shortly.[1]

The recently discovered "high-T_c" superconductor $Ba_{1-x}K_xBiO_3$ is unique in the sense that unlike the other high-T_c superconductors, it does not contain CuO_2 planes and is cubic rather than layered in structure. It does, however, exhibit a series of structural phase transitions as a function of both K concentration x and temperature. Superconductivity occurs only in a cubic phase and disappears abruptly at $x \leq 0.4$ where the behavior apparently becomes non-metallic.[2] Few details have been available hitherto of the structural phases found as x is decreased, and some of the reported results are not consistent with each other. The present measurements were carried out on the H4S triple axis spectrometer at Brookhaven's High Flux Reactor and on the Special Environment Powder Diffractometer (SEPD) at Argonne's Intense Pulsed Neutron Source.

The samples were prepared using the two-step synthesis method.[3,4,5] Pellets were made of stoichiometric mixtures of powder of Bi_2O_3, BaO and KO_2, fired in flowing nitrogen at a temperature below 725°C and then annealed in in flowing oxygen at a lower temperature. This procedure was repeated several times with intermediate grinding. The first batch of samples, with nominal compositions x = 0, 0.1, 0.2, 0.3 and 0.4 were run at Brookhaven. They were not completely single-phase and showed some oxygen deficiency, and the powder diffraction results were used primarily to obtain the structures of the different phases rather than to characterize the details of the phase diagram. A later synthesis produced more homogeneous samples of nominal compositions x = 0, 0.1, 0.2, 0.25, 0.3, 0.35, 0.4, 0.425, 0.5 and 0.6, which were studied at Argonne and used to refine the details of the structure and obtain the structural phase diagram. In both sets of experiments, the neutron powder diffraction (NPD) data were analyzed by full matrix Rietveld refinement.[6]

The structures which occur in the $Ba_{1-x}K_xBiO_3$ system can be viewed as distortions of the basic cubic perovskite structure. The fundamental distortions are rotations of nominally rigid BiO_6 octahedra or breathing distortions involving oxygen atom displacements towards or away from the Bi atoms, thus making the Bi atoms inequivalent. These distortions can occur singly or in combination to produce the various possible distorted perovskite structures. Glazer[7,8] has characterized the systematics of distortions produced by octahedral rotations about various axes, which serve to double the cubic perovskite unit cell. This

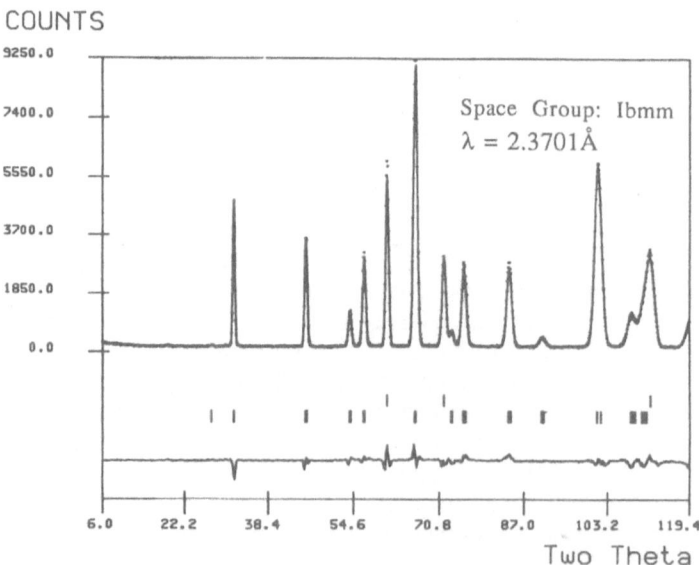

COUNTS

Space Group: Ibmm
$\lambda = 2.3701\text{Å}$

Two Theta

Figure 1(a)   Neutron powder diffraction data taken on $Ba_{0.9}K_{0.1}BiO_3$ at room
                temperature on the crystal diffractometer at Brookhaven, together
                with the Rietveld-refined fit. The lower curve represents the
                difference pattern between the measured and fitted profiles.

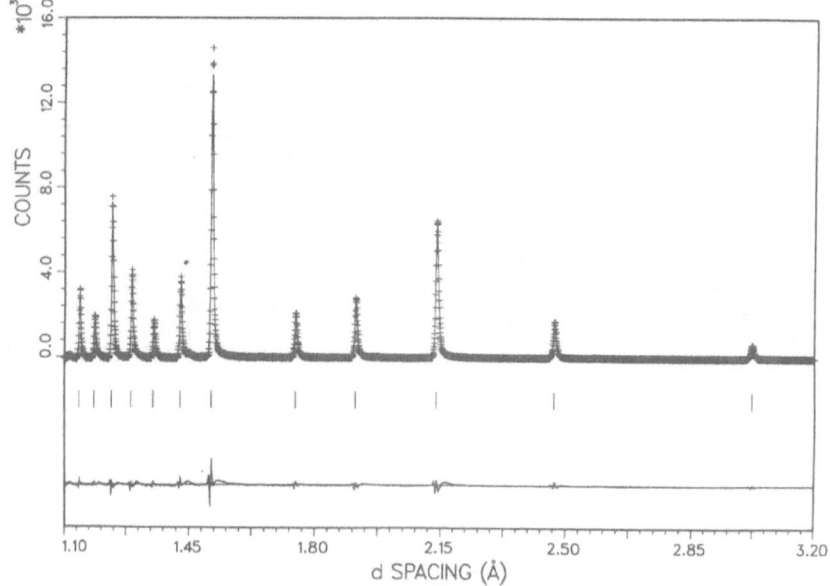

Figure 1(b)   Neutron powder diffraction data taken on $Ba_{0.6}K_{0.4}BiO_3$ at 10K on
the time-of-flight diffractometer at IPNS, Argonne, together with the
Rietveld-refined fit.

results in superlattice reflections at half-integer values of the cubic reflection
indices. These superlattice reflections are barely visible with x-ray diffraction,
owing to the weak scattering power of oxygen for x-rays, but can easily be seen
with neutron diffraction.

Examples of the neutron diffraction profiles and the Rietveld refinement fits
are shown in Figure 1(a) and (b). Figure 1(a) shows NPD data taken at room
temperature on a sample with nominal composition $Ba_{0.9}K_{0.1}BiO_3$. The
measurements were made on the triple-axis spectrometer at Brookhaven (set for
elastic scattering). The bars below the NPD profile indicate the positions of the
calculated Bragg reflections. (The upper set of bars represents the positions of
Bragg reflections from the Al container.) The lower curve represents the difference
pattern between observed and calculated diffraction profiles. The refinement is for
an orthorhombic structure (Space Group Ibmm). Figure 1(b) shows a time-of-
flight NPD profile taken at Argonne on $Ba_{0.6}K_{0.4}BiO_3$ at 10K fitted with a cubic
perovskite (Space Group Pm3m) structure. Figure 2 summarizes the phase
diagram obtained as a function of x and T. The room temperature $BaBiO_3$ (x = 0)
monoclinic structure I2/m agrees with earlier determinations of Thornton and
Jacobson[9] and Cox and Sleight[10,11]. At lower temperatures, a slightly better fit is
found refining to a primitive monoclinic structure, because of the appearance of new
weak superlattice reflections which could be indexed on the primitive monoclinic
cell. The high-temperature rhombohedral R3 phase was previously reported by
Cox and Sleight.[11].

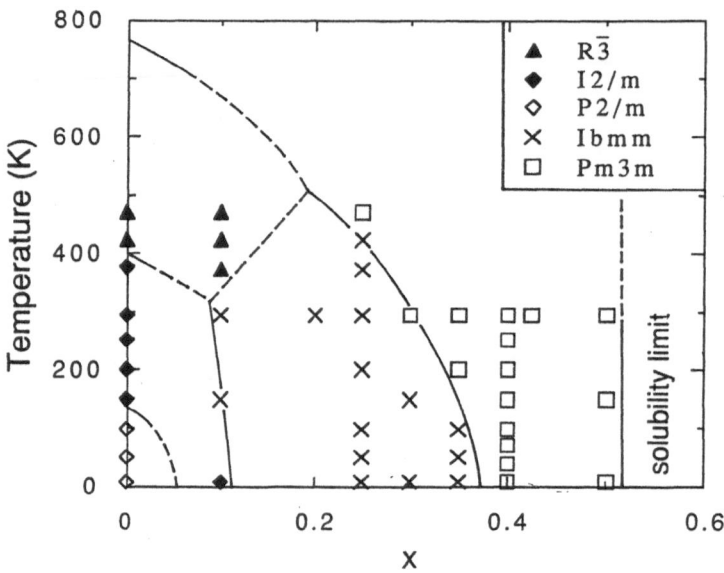

Figure 2   The structural phase diagram of the $Ba_{1-x}K_xBiO_3$ system as
                determined from the present measurements.

We now discuss the physical significance of the phase diagram shown in
Figure 2. If we follow the structures for increasing x at 200K, we see that we go
from a body-centered monoclinic phase (I2/m), characterized by an octahedral tilt
about the cubic [110] axis <u>coupled</u> with a breathing-mode distortion of the oxygen
displacements, to an orthorhombic phase (Ibmm), characterized by the [110] octahedral tilt only, with <u>no</u> breathing mode distortion, and finally to the cubic
perovskite superconducting phase (Pm3m) with no distortions. The actual
orthorhombic distortion of the unit cell should result in principle in a splitting of the
Bragg reflections. This was not resolved in the present experiments, as the
splitting of the a and b lattice constants is typically ~ 3 x 10^{-3} times their average
value. The high temperature rhombohedral R$\bar{3}$ phase for low x results from a
combination of a rhombohedral tilt about the [111] cubic axis <u>and</u> a breathing mode
distortion. The inequivalency of the Bi sites for the low x concentrations can be
regarded as the manifestation of a commensurate Charge Density Wave (CDW)
or charge disproportionation[9,10,11] of the Bi to Bi^{3+} and Bi^{5+}. Note that $BaBiO_3$
corresponds to the "half-filled band" limit where each Bi would have a single 6s
electron (nominal +4 valency) which is well known to be unstable due to electron
correlation effects. Unlike the similar half-filled band insulators in the cuprate
family of high-T_c superconductors, e.g., La_2CuO_4, $BaBiO_3$ apparently does not
become antiferromagnetic[12] but "solves" this instability problem via a CDW.[13]
The breathing mode distortion can produce an electron gap resulting in insulating

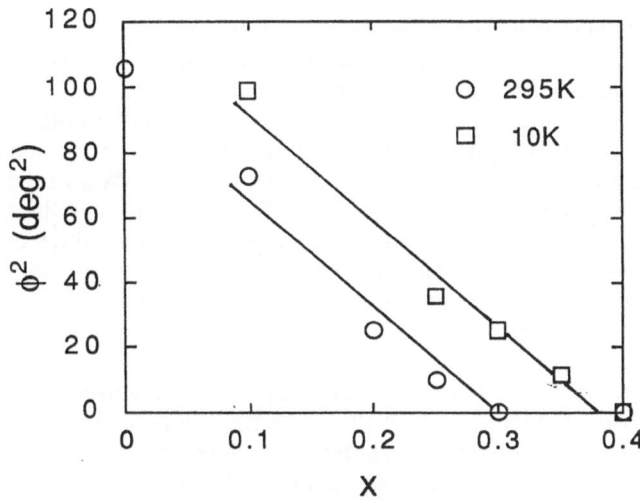

Figure 3.   The quantity ϕ^2 (ϕ = octahedral tilt) for the orthorhombic Ibmm phase versus potassium concentration, x, at 295K and at 10K.

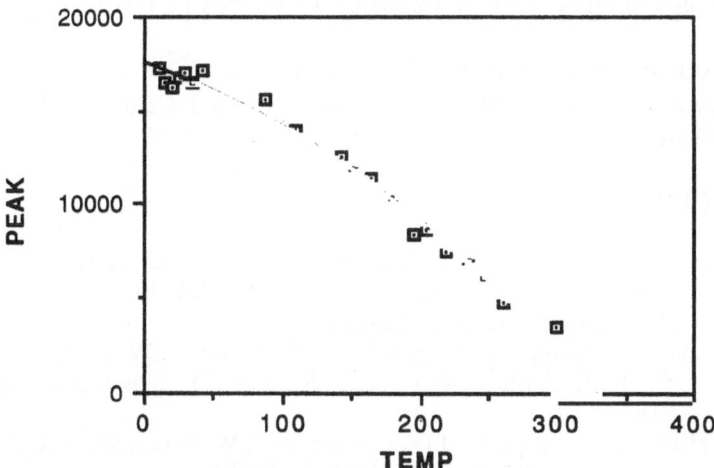

Figure 4.   The intensity of the (3/2 1/2 1/2) superlattice peak (in a cubic indexing scheme) versus temperature for a sample of nominal composition x = 0.3.

behavior. It is not, however, clear why the orthorhombic phase (characterized only by octahedral rotations but no CDW) and even the cubic phase (above T_c) continues to be insulating.[4] Figure 3 shows the square of the octahedral tilt angle versus x at 295K and 10K; it is seen to go continuously to zero at the orthorhombic/cubic transition. Schneemeyer et al.[14] have proposed an orthorhombic structure for small but finite x which includes a breathing-type distortion of the oxygen atoms (the Immm structure), but our NPD data show pronounced disagreement with such a structure. There have also been observations by electron diffraction of incommensurate CDW's in the whole x region up to the superconducting phase.[15,16] We could find no evidence in any of our NPD data for incommensurate superlattice peaks at any values of x or T. The conclusion is that such incommensurate modulations are fairly localized phenomena which do not exist throughout the bulk, or that they are in fact induced by the electron beam itself.

Thus, there are still several unresolved, puzzling features regarding the electronic behavior of this system in the insulating phases. Figure 4 shows the temperature dependence of the superlattice peak intensity of the orthorhombic phase which goes to zero continuously as the higher temperature cubic phase is approached. Figures 3 and 4 indicate that there should be a soft phonon mode in the vicinity of the orthorhombic-cubic phase transition, where T_c is in fact highest. Recent isotope effect measurements[17] and inelastic neutron scattering measurements[18] indicate that the superconductivity is due to electron-phonon interactions, although recent tunnelling measurements indicate that the electrons are more strongly coupled to relatively high frequency (~ 30 meV) phonons.[19]

The work at Argonne National Laboratory is supported by the U.S. DOE, BES-MS, contract No. W31-109-ENG-38, the NSF S&T Center for Superconductivity (SP and BD), and American Air Liquide, Inc. (DR).

REFERENCES

1. Shiyou Pei, J.D. Jorgensen, B. Dabrowski, D.G. Hinks, D.R. Richards, A.W. Mitchell, J.M. Newsam, S.K. Sinha, D. Vaknin, and A.J. Jacobson, Phys. Rev. B (to be published).

2. R.J. Cava, B. Batlogg, J.J. Krajewski, R.C. Farrow, L.W. Rupp, Jr., A.E. White, K.T. Short, W.F. Peck, Jr., and T.Y. Kometani, Nature 332: 814 (1988).

3. D.G. Hinks, B. Dabrowski, J.D. Jorgensen, A.W. Mitchell, D.R. Richards, S. Pei, and D. Shi, Nature 333: 836 (1988).

4. B. Dabrowski, D.G. Hinks, J.D. Jorgensen, R.K. Kalia, P. Vashishta, D.R. Richards, D.T. Marx, and A.W. Mitchell, Physica C156: 24 (1988).

5. D.G. Hinks, D.R. Richards, B. Dabrowski, A.W. Mitchell, J.D. Jorgensen, and D.T. Marx, Physica C156: 477 (1988).

6. H.M. Rietveld, J. Appl. Cryst. 2: 65 (1969); R. B. Von Dreele, J. D.

Jorgensen, and C. G. Windsor, J. Appl. Cryst. 15: 581 (1982)
7. A.M. Glazer, Acta Cryst. B28: 3384 (1972).
8. A.M. Glazer, Acta Cryst. A31: 756 (1975).
9. G. Thornton and A.J. Jacobson, Mater. Res. Bull. 11: 837 (1976).
10. D.E. Cox and A.W. Sleight, Solid State Comm. 19: 969 (1976).
11. D.E. Cox and A. W. Sleight, Acta Cryst. B35: 1 (1979).
12. Y.J. Uemura et al., Nature 335: 151 (1988).
13. C.M. Varma, Phys. Rev. Lett. 61: 2713 (1988).
14. L.F. Schneemeyer, J.K. Thomas, T. Siegrist, B. Batlogg, L.W. Rupp, R.L.
    Opila, R.J. Cava, and D.W. Murphy, Nature 335: 421 (1988).
15. S. Pei, N.J. Zaluzec, J.D. Jorgensen, B. Dabrowski, D.G. Hinks, A.W.
    Mitchell, and D.R. Richards, Phys. Rev. B 39: 811 (1989).
16. E.A. Hewat, C. Chaillout, M. Godinho, M.F. Gorius, and M. Marezio,
    Physica C157: 228 (1989)
17. D.G. Hinks, D.R. Richards, B. Dabrowski, D.T. Marx, and A.W. Mitchell,
    Nature 335: 419 (1988).
18. C.K. Loong, P. Vashishta, R.K. Kalia, M.H. Degani, D.L. Price, J.D.
    Jorgensen, D.G. Hinks, B. Dabrowski, A.W. Mitchell, D.R.
    Richards, and Y. Zhang, Phys. Rev. Lett. 62: 2628 (1989).
19. J.F. Zasadzinski, N. Tralshawala, D.G. Hinks, B. Dabrowski, A.W.
    Mitchell, and D.R. Richards, Physica C 158: 519 (1989)

ELECTRON MICROSCOPIC STUDIES OF OXYGEN ORDERING PHENOMENA IN $YBa_2Cu_3O_{7-\delta}$

S. Amelinckx, G. Van Tendeloo, and
J. Van Landuyt

University of Antwerp (RUCA)
Groenborgerlaan 171, B-2020 Antwerp
Belgium

ABSTRACT

The order-disorder phemomena related to the oxygen sublattice in the $CuO_{1-\delta}$ layers of $YBa_2Cu_3O_{7-\delta}$ are studied by means of electron diffraction and electron microscopy. Twinning as associated with the orthorhombic to tetragonal phase transition are described in detail. Different superstructures, due to oxygen vacancy ordering are analysed. In constant stoichiometry cooled samples the OII phase is found in a wide composition range; it is long ordered also along the c direction. Its spatial structure is described. The $2\sqrt{2}a_o \times 2\sqrt{2}a_o$ structure is shown to be presumably due to deformation modulation rather than to oxygen ordering. The results are compared with the theoretical predictions based on an Ising model.

1. INTRODUCTION

Oxygen stoichiometry and oxygen order-disorder phenomena have been identified as important ingredients of superconductivity in oxides and in particular in $YBa_2Cu_3O_{7-\delta}$. In the present survey, largely based on ref. [1][2][3] we shall illustrate the contribution of electron microscopy and electron diffraction to the study of these phenomena. Although the scattering of electrons by

9

$Ba_2Y Cu_3 O_{7-\delta}$

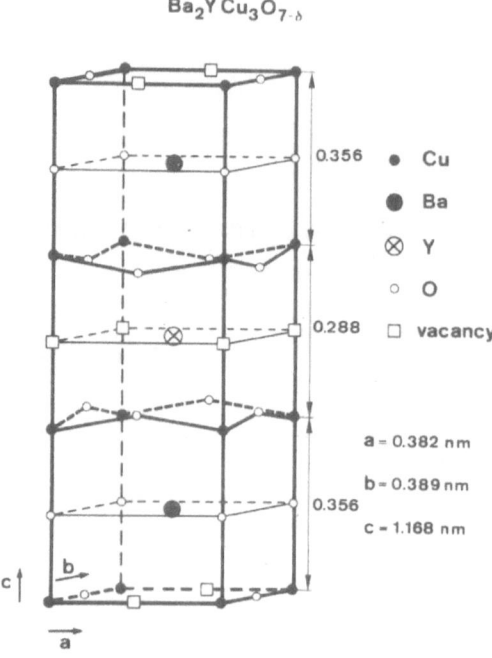

• Cu

● Ba

⊗ Y

○ O

□ vacancy

0.356

0.288

0.356

a = 0.382 nm

b = 0.389 nm

c = 1.168 nm

Fig. 1 Schematic representation of the $YBa_2Cu_3O_7$ structure.

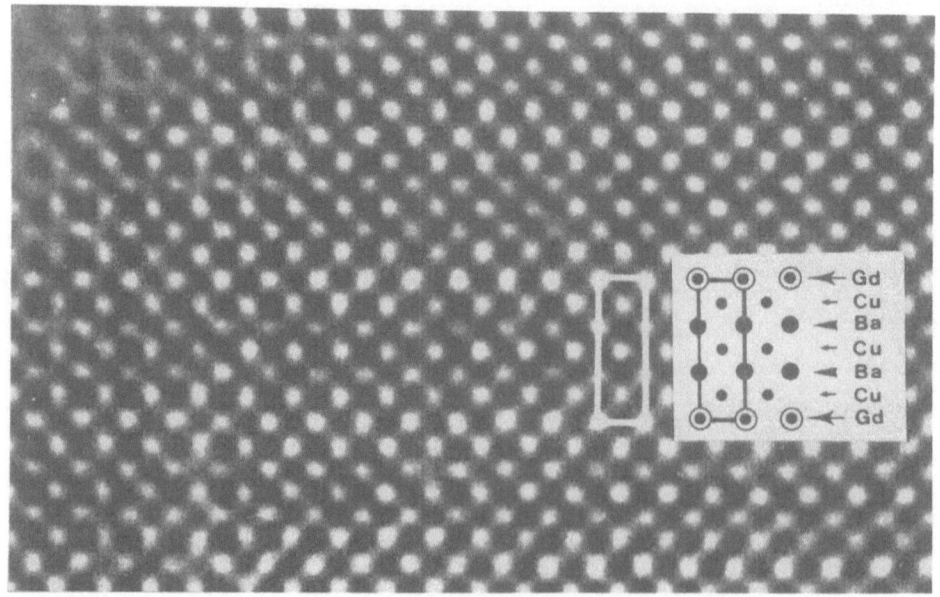

Fig. 2 High resolution image of $GdBa_2Cu_3O_7$ as viewed along [010] or [100].

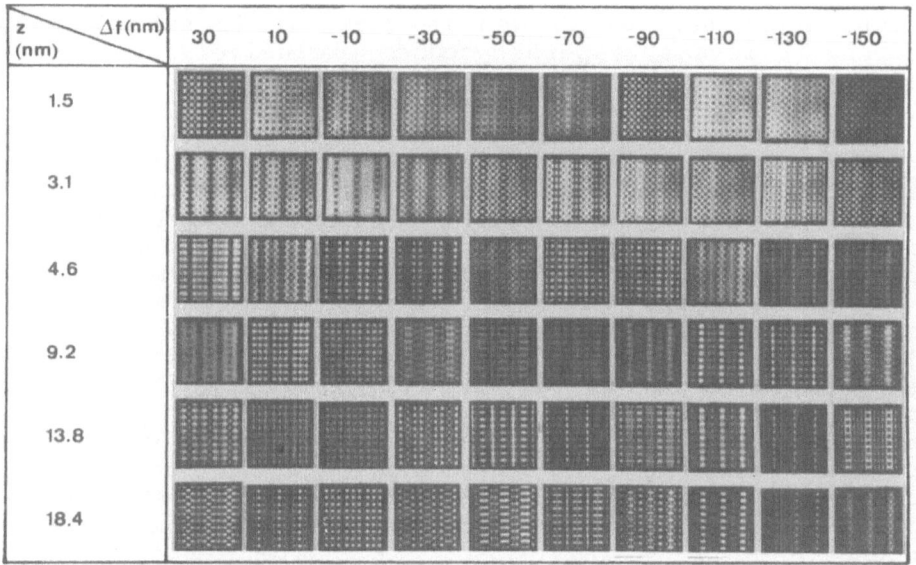

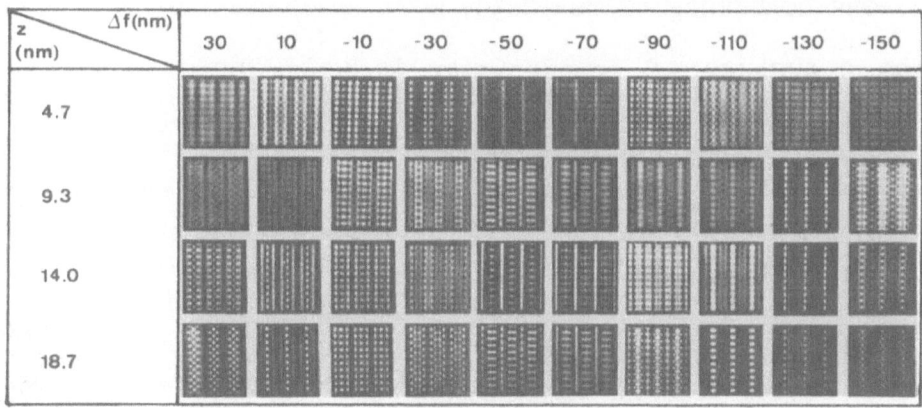

Fig. 3 Matrices of computer generated images of
$YBa_2Cu_3O_7$ (t = thickness, Δf = defocus).
(a) along the [100] zone
(b) along the [010] zone

whereas in the copper layers, situated between the two
BaO layers, half of the available oxygen sites remain
unoccupied. The oxygen atoms in this layer are arranged
in rows along the b_o direction so as to form Cu-O-Cu-O-
... chains; this is the only feature of the structure
that breaks the tetragonal symmetry. The unoccupied oxygen sites can be considered as "structural vacancies"

light atoms such as oxygen is weak, it is sufficiently
strong to reveal superstructures exclusively due to the
ordering of oxygen.

High resolution electron microscopy and electron
diffraction have made important contributions to the
study of the structural features of superconductors, because they make it possible to obtain single crystal
diffraction patterns of very small crystal fragments and
allow moreover to form an atomic resolution image of the
same small crystal area. Such experimental images, especially when their interpretation is supported by computer simulated images based on proposed structural models, allow to obtain very rapidly a first approximation
of the structure of new complex compounds. Also they
have been of considerable help in suggesting new compounds because they allow to detect local stacking sequences, deviating from the expected one, hereby suggesting structures and compositions of possible new compounds.

2. THE STRUCTURE OF $YBa_2Cu_3O_{7-\delta}$

The structure of $YBa_2Cu_3O_{7-\delta}$. is well known [4][5]. A
schematic view is represented in Fig.1. We shall draw
the attention to those features which are of importance
for the subsequent discussion. At room temperature the
stoichiometric compound (i.e. for $\delta = 0$) is orthorhombic, pseudo-tetragonal with lattice parameters $a_o \cong b_o$
and $c_o \cong 3a_o$ ($a_o = 0.382$ nm; $b_o = 0.389$ nm and $c_o = 1.168$
nm). The spacegroup is Pmmm and the pointgroup 2/mmm.
The structure is a defective derivative of the cubic
perovskite structure ABO_3. The unit cell contains three
perovskite cubes, juxtaposed along the c-direction. The
summits of the cubes are occupied by copper atoms and
their centres by yttrium and barium in the succession
Ba-Y-Ba-Ba-Y-... The succession of layers along the c-direction is thus

$$CuO-BaO-CuO_2-Y-CuO_2-BaO-$$

In the corresponding stoichiometric perovskite the succession of layers would be

$$CuO_2-BaO-CuO_2-YO-CuO_2-BaO$$

and the composition $(Ba,Y)_3Cu_3O_9$. We note that in the superconducting compound the Y layer is free of oxygen,

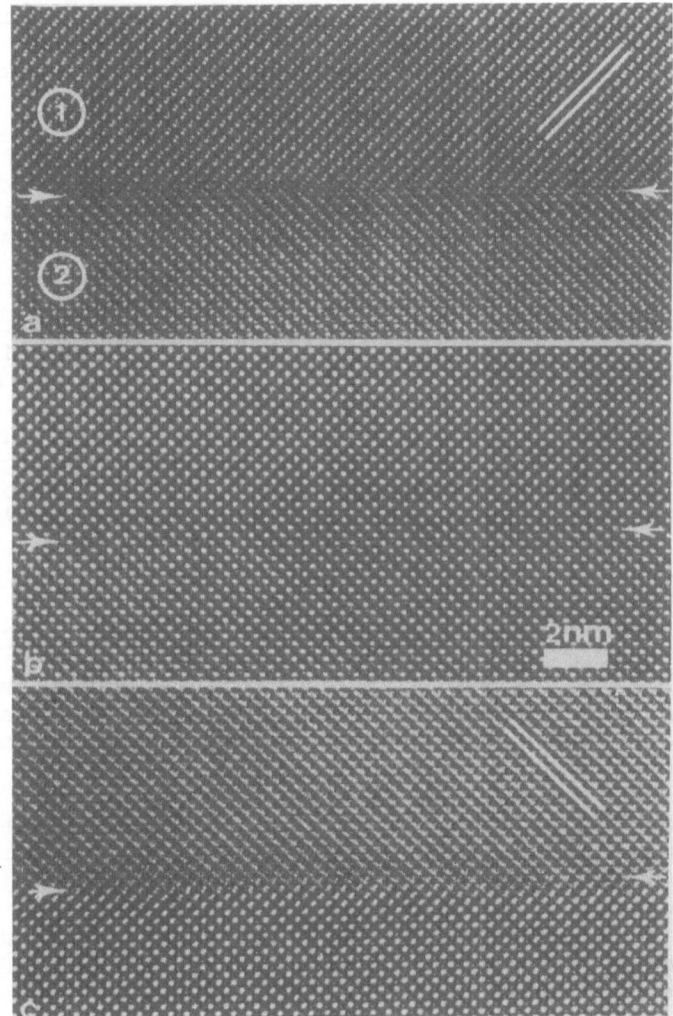

Fig. 4 High resolution image along [001] of a twin interface indicated by an arrow. The three images
refer to the same area; they only differ in defocus.

i.e. vacancies which result from the fact that the
structure offers more equivalent sites than can actually
be occupied by the available atoms. Such vacancies make
diffusion in this layer easy since no vacancy formation
energy must be supplied. If the oxygen atoms (or the vacancies) in the CuO layer are disordered the structure
becomes tetragonal; this occurs at a sufficiently high
temperature. The structure also becomes tetragonal if δ
is sufficiently large; up to $\delta \sim 0.6$ the structure remains orthorhombic at room temperature. However whether

a material with a given composition is orthorhombic or tetragonal at room temperature may also depend on the degree of order i.e. on the heat treatment.

3. IMAGING OF THE PERFECT STRUCTURE

Fig.2 shows a high resolution image of $GdBa_2Cu_3O_{7-\delta}$,the inset indicates its relation with the structure model. It is clear that the image is consistent with the structure of Fig.1.

Matrices of simulated images along different crystallographic directions: [100],[010] and [001] and for different specimen thicknesses and defocus values are reproduced in Fig.3. The differences between the images along the [100] and [010] zones are very small. The same types of images are found in the matrices of Fig.3a and Fig.3b i.e. in the [100] and [010] views. Moreover the images are sensitively dependent on the thickness and defocus which are usually unknown but have to be determined from the images themselves. It is therefore very difficult to say whether for instance Fig.2 is a [100] or a [010] view. The similarity is of course due to the fact that only the oxygen arrangement in one layer per unit cell is different in the two views. Nevertheless it is possible to reveal the effect of the oxygen arrangement in a [001] zone image because along this zone the two orientations occur in adjacent regions of a twinned crystal and can thus be imaged under identical conditions. The [001] images of Fig.4b were made under these conditions; they are from the same area but correspond with different defocus values. At exact Scherzer focus (Fig.4a,c) they exhibit pseudo square arrays of sharp circular dots; on under- or over focussing the dots become somewhat elongated along the chain directions or perpendicular to this direction, forming dotted lines (Fig.4). The simulated images of Fig.5 predict precisely such a behaviour. We conclude that an observable effect of the oxygen ordering can thus be expected in electron microscopic images made under adequate conditions.

4.TWINNING

The simplest and most striking manifestation of the ordering of the oxygen atoms (or vacancies) in the $CuO_{1-\delta}$ layers is the occurrence of twinning in the orthorhombic (OI) phase.

Diffraction contrast images of OI specimens viewed along [001] almost invariably exhibit bands limited either by (110) or by ($1\bar{1}0$) planes and which have differ-

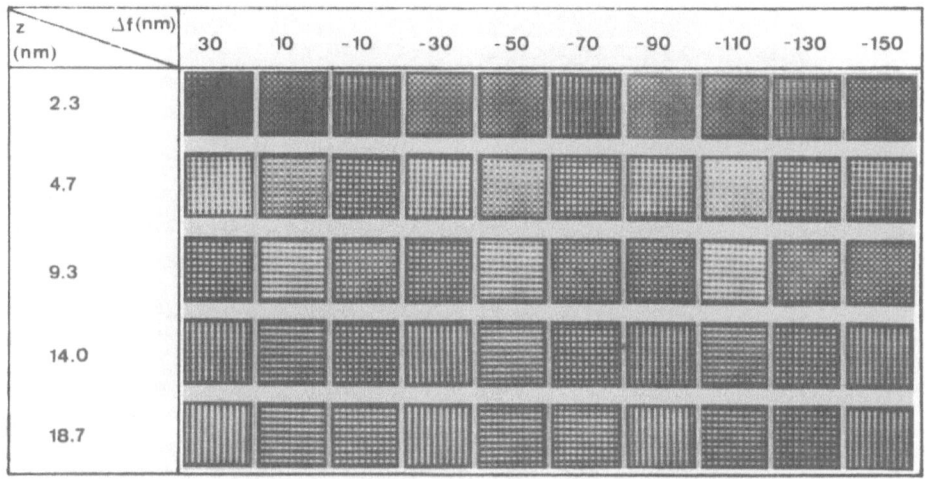

Fig. 5 Matrix of computer simulated images viewed along
[001] (cfr. Fig.4) (t = thickness, Δf = defocus).

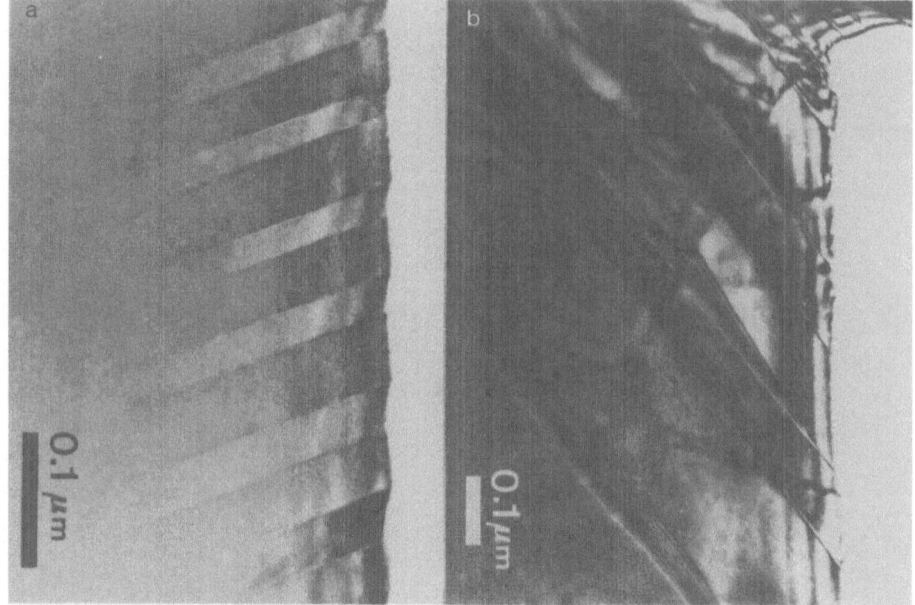

Fig. 6 Twin bands in $YBa_2Cu_3O_{7-\delta}$ as imaged in diffraction
contrast
(a) domain contrast
(b) interface contrast

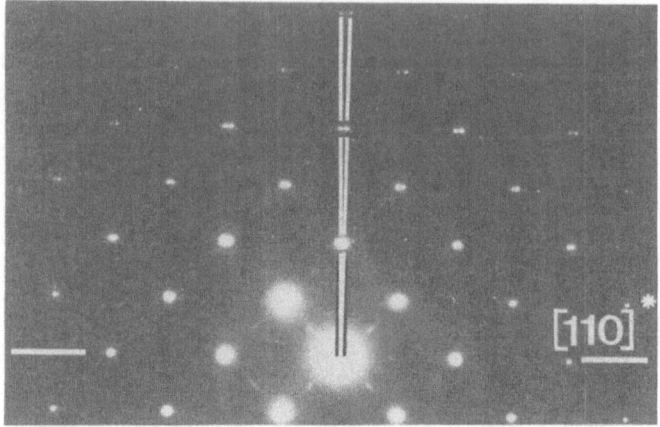

Fig. 7 Diffraction pattern along [001]* across a twin
       interface. The row of unsplit spots is along
       [110]*; all other spots are split.

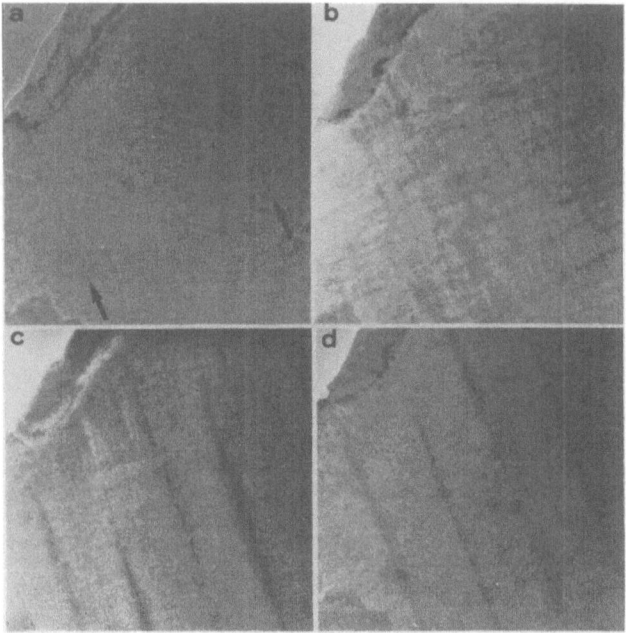

Fig. 8 Successive stages of a heatpulsing experiment
       (a) initial configuration (b) immediately after
       heatpulsing (c) fine twins are reformed; the in
       terfaces are vague (d) wide twins have been re-
       formed

ent shades under most diffraction conditions (Fig.6).
The contrast effects prove that these specimen areas
differ in orientation. A diffraction pattern along
[001] from an area crossing the contact plane between
two such regions is shown in Fig.7. The pattern exhibits
a central line of unsplit spots parallel to [110]*, all
other spots being split along the direction parallel
with the unsplit row the magnitude of the splitting
being proportional with the distance from the unsplit
row. Such a pattern is the signature of a reflection
twin, the symmetry plane being perpendicular to the
unsplit row. Since the habit plane is also the (110)
plane, the twins are coherent.

The twinning results from the fact that the crystal
is formed at high temperature where the oxygen atoms in
the CuO layers are disordered and as a result, the
structure is tetragonal with point group 4/m 2/m 2/m of
order 16. On lowering the temperature the structure becomes orthorhombic with pointgroup 2/m 2/m 2/m of order
8. The number of orientation variants is thus 16:8=2.
These orientation variants are related by symmetry elements lost during the transformation,for instance by
the (110) or (1$\bar{1}$0) mirrors. Alternatively one may choose
as a symmetry operation a rotation over 90° about the
fourfold axis. Whatever the choice of the symmetry element the a_o and b_o axis in the two parts of the twins
are interchanged. The strain free interfaces between the
two different orthorhombic variants are the (110) and
(1$\bar{1}$0) planes; they are entirely determined by symmetry
[6]. The observed configurations of twins minimize the
strain energy resulting from the T $\longrightarrow$ O transformation.
As shown above the orthorhombicity of the structure is
directly associated with the ordering of the oxygen
atoms in chains along the b_o direction. We thus conclude
that the phase transformation T $\longrightarrow$ O must be due, at
least in part, to a disorder-order transformation on the
oxygen sublattice of the CuO layers.

The phase transformation has been studied "in-situ"
in the electron microscope, using the electron beam as a
heat source. On heating the twin interfaces are found to
disappear at a temperature $\sim$ 700°C depending on the composition of the sample. They reappear on cooling. This
operation can be repeated a number of times provided the
sample is kept only a short time just above the temperature required to cause the transformation.Successive
stages of this experiment are reproduced in Fig.8 and
were recorded on a video recording [7].

Immediately after heatpulsing the specimen exhibits a
"tweed" texture i.e. two mutually perpendicular line
systems which gradually transform into two families of
fine twin lamellae. The twin interfaces do not reappear
at the same positions as the ones they occupied before
the transformation i.e. there is no "memory effect".
Nevertheless in any given area the same family of twin
interfaces tends to reappear prominently after each successive transformation cycle.The reformation of twins is
much slower than their disappearance. A short time (1
min) after heatpulsing the twins are fine and the interfaces are ill defined, they coarsen gradually in a matter of 10-20 min when keeping the specimen at normal microscope temperature and the interfaces become sharper.
The rate of widening of the twin lamellae is consistent
with a rapid diffusion controlled process. The video
tape recording of the process has revealed that towards
the end of the coarsening process, widening also occurs
by the rapid propagation of ledges along the twin interfaces, suggesting also a diffusionless component in the
transformation.

The same process can be followed via the diffraction
pattern (Fig.9). We start with a DP exhibiting clearly
the spot splitting due to the orthorhombic twins. Immediately after heat pulsing the diffraction pattern looks
tetragonal i.e. there is no spot splitting but the spots
form small crosses, the arms of the cross pointing along
the $[1\bar{1}0]^*$ and $[110]^*$ directions. Also diffuse streaks
passing through the rows of spots are usually visible.
These features can be observed in Fig.9. After 10-20
min the diffraction spots are again split, the magnitude
of the splitting being possibly somewhat smaller than
the original one. These observations can consistently be
interpreted in terms of the disordering of the oxygen
atoms by the temperature pulse. Already immediately after the end of the pulse the crystal consists of very
small microdomains each containing chains with a preferential orientation along one out of the two possible directions. These microdomains are in fact already very
fine orthorhombic twins limited by (110) and $(1\bar{1}0)$ as
suggested by the cross-like shape of the diffraction
spots. The average macroscopic symmetry as revealed by
the DP is then tetragonal, but the shape of the spots
reveals the presence of the microtwins i.e. the local
microscopic, symmetry is already orthorhombic. The subsequent widening of the twins is a direct consequence of
the rapid diffusion of oxygen in the $CuO_{1-\delta}$ layers, which
is made possible by the presence of the structural vacancies in these layers. The sideways migration of the
twin interfaces implies the cooperative jumping of oxy-

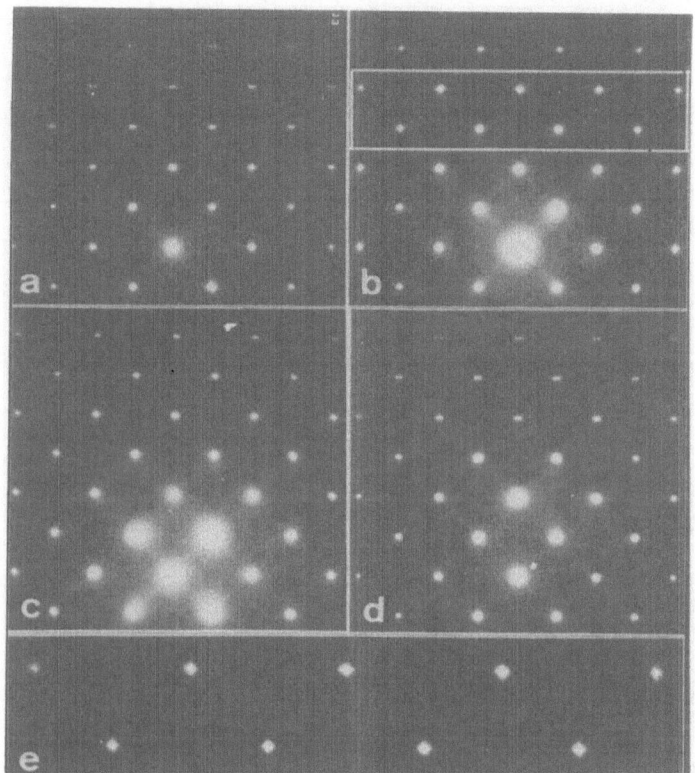

Fig. 9 Diffraction patterns corresponding with the different stages of Fig.8
    (a) initial situation; the crystal is
        orthorhombic
    (b) the diffraction spots form small crosses. The
        macroscopic symmetry is tetragonal
    (c) spot splitting begins to reappear
    (d) final situation: the spot splitting is recov
        ered almost completely
    (e) magnified view of (b)

gen atoms from sites along rows having one orientation
into sites belonging to rows having the other orientation. Provided the heat pulse was short enough not to
loose a significant amount of oxygen into the microscope
vacuum, the process is reversible and the same spot
splitting will ultimately be recovered. If loss of oxygen has occurred the final spot splitting will be
smaller than the initial one; if too much oxygen is
lost the structure may even remain tetragonal and the
final composition may become $YBa_2Cu_3O_6$.

The video tape recording [7] of the T $\longrightarrow$ O transformation process   suggests a diffusionless component in the widening process, which occurs when the twins have already a significant width. This can be understood by noting that the ordering of the oxygen atoms causes an orthorhombic deformation, leading in turn to the building up of "transformation stresses" which may become large enough to nucleate and cause the propagation of twinning dislocations along the interface.

In well-annealed specimens the twin interfaces are sharply defined, the atomic rows remain straight up to the twin interface (Fig.4). At the interface the lattice abruptly changes orientation over an angle ϕ

$$\phi = 2 \arctan(b_o/a_o) - \pi/2$$

which is of the order of 1.04°. The same angle determines the spot splitting in the diffraction pattern.

The orthorhombicity being only due to the chain formation along b_o, the twinning is a consequence of the change in orientation of these chains along the interface. An idealized model of the twin boundary thus consists of a continuous copper sublattice filled with oxygen along mutually perpendicular directions (Fig.10). Presumably some rearrangement of the atoms will take place along the interface so as to create a coordination for the copper atoms in the boundary that better approximates the coordination in the bulk. Such rearrangements might cause some lattice relaxation along the boundary i.e. the separation of the (110) lattice planes parallel with the boundary plane might be slightly different close to the interface from that in the bulk. As a result the lattices in the two parts would be slightly shifted one with respect to the other. Image treatment of high quality high resolution images of boundaries viewed along [001] has not revealed such an effect as yet, putting an upper limit of about 1% of the lattice parameter to this relative shift.

5. VACANCY ORDERED STRUCTURES

5.1 <u>The 2a_o structure in beam heated specimens</u>

The interpretation of the observations on twinning has shown that structural vacancies order below a certain critical temperature in $YBa_2Cu_3O_7$. It is therefore to be expected that the additional structural vacancies in non-stoichiometric $YBa_2Cu_3O_{7-\delta}$ ($\delta > 0$) will form or

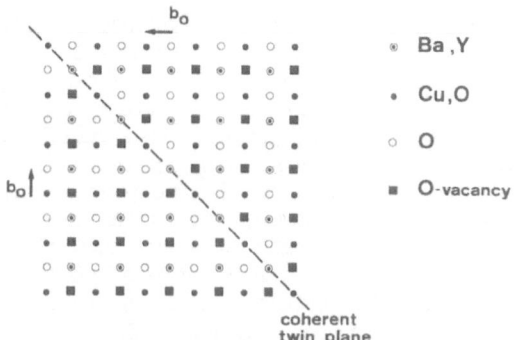

Fig. 10 Idealized model for a coherent twin boundary.

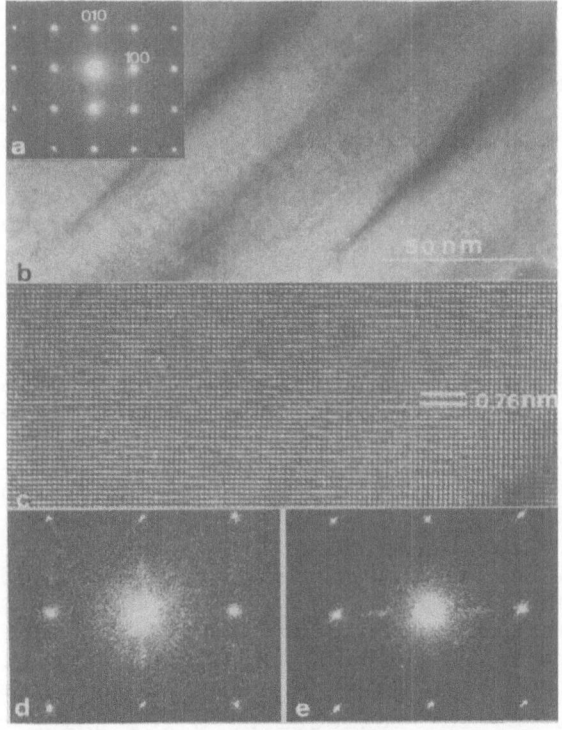

Fig.11 (a) diffraction pattern of specimen exhibiting the $2a_0$ structure (b) low magnification high resolution image of the selected area; two orientation variants of the $2a_0$ structure are present.(c) high magnification of (b); note the doubling of the spacing between bright dot rows(d) and (e) optical diffraction pattern of (b); the two orientation variants are separated.

dered arrangements as well. Evidence that this indeed
happens was first found in specimens which had been
heated in the microscope []. After a number of
heatpulses (in the microscope vacuum) the specimen
mostly becomes oxygen deficient. The diffraction pattern
of such specimens then looks as in Fig.11 i.e. weak and
diffuse additional spots are formed at position $h+1/2$ 0
0 and 0 $k+1/2$ 0, whilst the spot splitting is still present i.e. the crystal is still orthorhombic. The corresponding high resolution image then reveals weak extra
bright dot rows parallel with the b_0 direction and with
a spacing $2a_0$. This image shows furthermore that the
diffraction pattern is the superposition of two diffraction patterns due to two orientation variants of a superstructure, with unit mesh $2a_0 \times b_0$, inherited from the
two orientation variants of the orthorhombic I structure.

From the [010] section of reciprocal space we conclude that the $h+1/2$ 0 1 spot rows are degenerated into
weak continuous streaks. The simplest, somewhat idealized model consistent with the observed unit mesh, and
assuming that the ordering pattern is confined to the
CuO layers, is shown in Fig.12b, it corresponds to a
composition with $\delta = 0.5$.

Computer simulated images of the $2a_0$ structure based on
this model are reproduced in Fig.12. The contrast of the
extra bright dots is weak especially in the thin parts
of the foils (Fig.12b right) used in high resolution,
but it increases in thick specimens (Fig.12b left). The
observations (Fig.11c) also show a weak contrast, and
are consistent with the computed images. In this model
the Cu-O-Cu-O chains are separated by $2a_0$. The $2a_0$
arrangements can be formed in two different positions
two translation variants are possibly related by
displacement $\mathbf{R} = [100]$. The diffuse streaks along c^*
suggest that there is no correlation between the translation variants present in successive layers. The fault
vector of the stacking is $\mathbf{R} = [100]$, when referred to
the OI lattice. The reflections $\mathbf{g}$ for which $\mathbf{g.R} \neq$
integer will then degenerate into streaks i.e. the h 0 1
rows with fractional indices h will become streaked
whilst the spots in the other h 0 1 rows remain sharp,
as observed. The presence of two translation variants
also leads to the formation of anti-phase boundaries
(Fig.13). The width of the 1/2 0 0 spots in the a_0 direction suggests the presence of some disorder in this
direction. In some cases the position of the peak of the

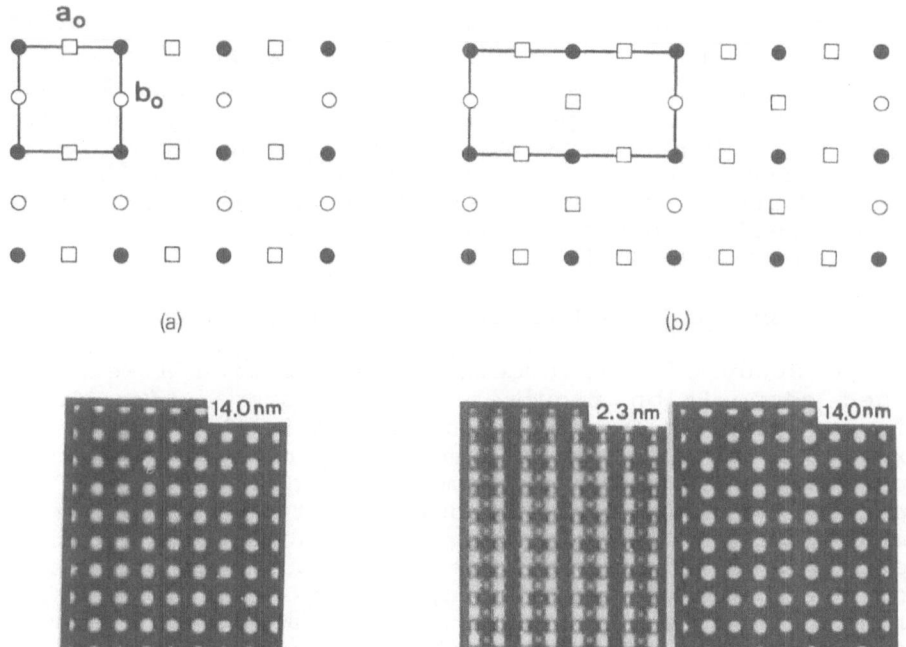

Fig.12 Models of the CuO layers in the OI and OII structures : (a) OI structure; model and high
resolution image along [001]. (b) OII structure;
model and high resolution image along [001]
for two different thicknesses but for the same
defocus as (a)

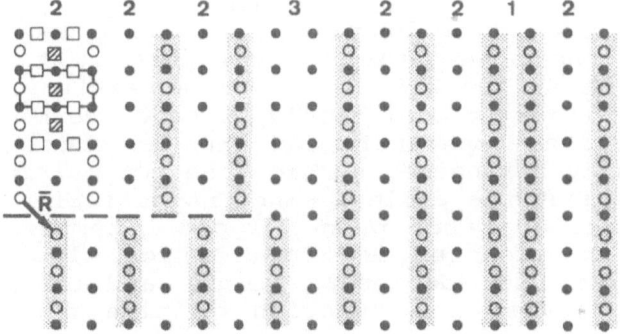

Fig.13 Model for the defective $2a_0$ structure. Note the
variability in spacing of the chains, and the
presence of anti-phase boundaries.

Lorentzian distribution is not exactly at 1/2 0 0 but at
1/2 - ε 0 0. This can be understood by assuming that the
interchain spacing is a mixture of $2a_0$, a_0 and $3a_0$, or
perhaps even other spacings. The disorder in the beam
treated specimens is essentially due to the fast cooling
of such specimens under poorly controlled conditions. We
shall further discuss observations on specimens made
from materials prepared under stringent conditions of
composition and cooling rate.

5.2 Constant stoichiometry cooled material (CSC)

The study of well ordered material with a well defined composition requires special preparation techniques which we now discuss. In early investigations
oxygen ordering phenomena have mainly been studied using
specimens quenched from some temperature whereby the
specimen is equilibrated under a certain partial pressure of oxygen. The quenching treatment is used in order
to obtain at room temperature the composition present at
the equilibration temperature, but it results in poorly
ordered specimens. Cooling slowly in a gaseous environment with constant composition results in materials
with gradients in the oxygen concentration and hence
with not well-defined composition. It is thus required
to "uncouple" cooling rate and composition. This can be
achieved by the use of the constant stoichiometry cooling technique which was recently developed by Verwey and
Bruggink [8]. This method consists in monitoring continuously the specimen's weight and the oxygen partial
pressure over the specimen. The latter is adapted continuously by means of a feed-back system in such a way
that the weight of the specimen is kept constant on
cooling. This method ensures that the composition obtained at the observation temperature is the same as
that present at the equilibration temperature, but nevertheless allows to cool slowly through the disorder-order transition temperature and thus to obtain well ordered vacancy arrangements. Also the composition is uniform throughout the pellets since no diffusion in or out
of the specimen takes place. Typical temperature-time
cycli are represented in Fig.14 which also plots the
variation of the specimen's weight and of the oxygen
partial pressure as a function of time for a typical
run.

Microscopy specimens are obtained by grinding the
pellets and depositing the fine crystal fragments on a
holey film or glueing them on copper grids; this operation does not change the composition and involves no
heat treatment.

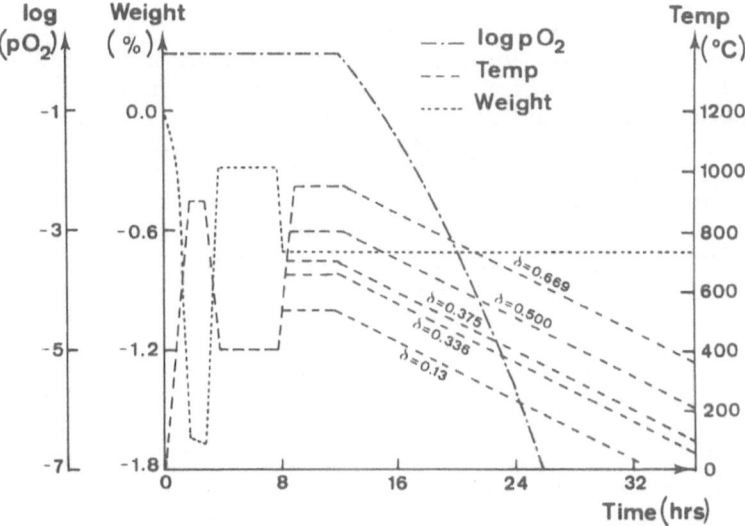

Fig.14 Temperature-time program applied in the constant stoichiometry cooling method for different δ values. Also the specimen's weight and the oxygen partial pressure is plotted for one run (δ = 0.2, T_{eq} = 600°C).

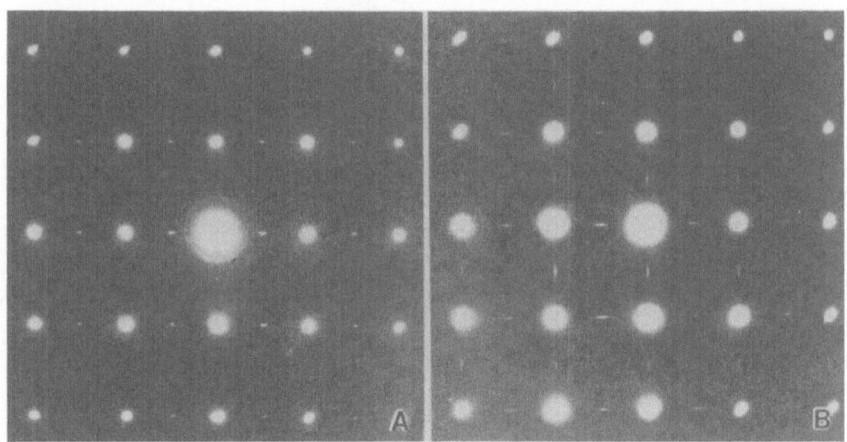

Fig.15 Electron diffraction patterns along the [001]* zone of a constant stoichiometry cooled specimen ($\delta \sim$ 0.5) : (a) single variant; (b) two orientation variants

The materials prepared by the CSC—method have made it possible to perform a systematic study of the vacancy ordered phases. In material with $\delta \cong 0.5$ one finds predominantly the $2a_0$ superstructure, described above. However whereas in beam treated specimens the superstructure spots were rather weak, diffuse and elongated along the $a_0{}^*$ direction, we now find quite sharp and relatively intense spots at the positions $h+1/2$ k 0 in the [001] zone (Fig.15). Also we could now isolate relatively large domains containing a single variant and producing a well defined diffraction pattern such as the one shown in Fig.15a. This confirms our interpretation of the diffraction pattern with spots at $1/2$ 0 0, and 0 $1/2$ 0 as being due to two orthorhombic twin variants in the selected area. Furthermore also the [010] zone diffraction patterns now reveal sharp spots at positions $h+1/2$ 0 1 (Fig.16), compared to the continuous streaks found in quenched specimens. The c—parameter of the $2a_0$ (or OII) structure is thus found to be the same as that of the OI structure i.e. $c \cong 3a_0$. The sharpness of the spots indicates that the vacancy configurations in successive $CuO_{1-\delta}$ layers are now well correlated. A complete spatial model of the structure can now be proposed, it is represented in Fig.17a. Only if the vacancy arrangements in successive $CuO_{1-\delta}$ layers are stacked vertically above one another is the c—parameter equal to $3a_0$.

Fig.18c shows the high resolution image along [001] of a constant stoichiometry specimen with $\delta = 0.5$, made under conditions of strong defocussing in a relatively thick specimen area so as to emphasize the superperiod rather than the atomic resolution. The extension of the areas exhibiting the $2a_0$ period can be judged from dark field images made in a single $1/2$ 0 0 superstructure spot, such as the one reproduced in Fig.18b. The $2a_0$ areas show up as bright patches; only one variant is revealed. Under 200 or 400 kV electron irradiation the superperiod disappears gradually within minutes, which is consistent with the assumption that it is due to the ordering of light atoms.

High resolution images along the [010] zone (Fig.16) confirm the model of Fig.17a; this image also exhibits the presence of anti-phase boundaries with a displacement vector $\mathbf{R} = 1/2[100]$, when referred to the $2a_0 \times b_0 \times c_0$ unit cell of the OII phase.

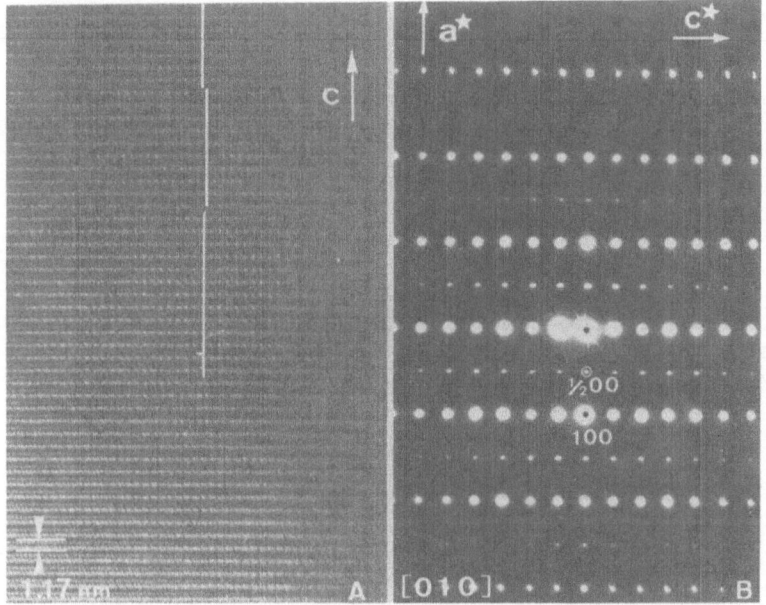

Fig.16 (a) high resolution image along the [010]* zone:
presence of anti-phase boundaries.(b) electron
diffraction pattern along the [010]* zone of a
constant stoichiometry cooled specimen ($\delta \sim 0.5$)

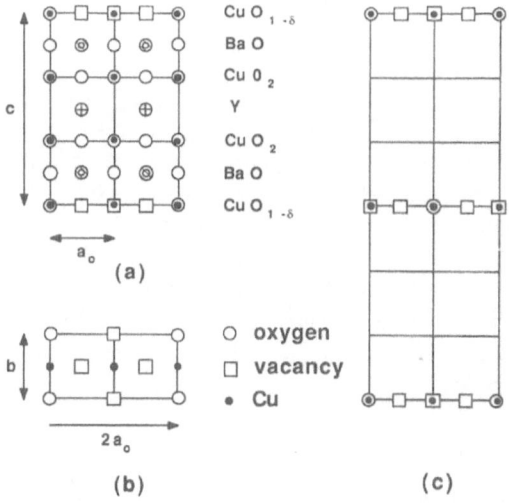

(a)

(b)

(c)

Fig.17 Spatial models
of the O_{II} structure
(a) the most fre-
quently occuring
structure; the con-
figurations in suc-
cessive $CuO_{1-\delta}$ layers
are vertically
stacked
(b) the arrangement
in the $CuO_{1-\delta}$ layers
(c) alternative model
with a doubled c
parameter occurring
occasionally

In certain specimens one observes small regions in which the stacking is predominantly as represented in Fig.17c i.e. successive $CuO_{1-\delta}$ layers are shifted over $a_o[100]_I$ (or $1/2a_o[100]_{II}$), depending on the reference unit cell. The resulting arrangement as viewed along the [010] zone is then centered; this is visible in the high resolution image of Fig.19c. Such regions are too small to produce an electron diffraction patttern with well defined spots, but optical diffraction of the high resolution image of such a region reveals a centered lattice (Fig.19b). The energy difference between the two stacking modes must be small; this is the reason why different stacking modes are likely to occur either periodically (polytypism) or at random. In the latter case the diffraction patterns will show h+1/2 0 l rows which are streaked.

5.3 The na_o structures

In CSC specimen with $\delta \cong 1/3$ one finds occasionally evidence in the [001] zone for a tripling of the a_o parameter. The diffraction pattern is reproduced in Fig.20; the superstructure spots are somewhat diffuse and elongated along a_o but their peaks divide the distance 000-100 in three almost exactly equal parts; two orientation variants are present. Idealized models for the $3a_o$ configuration in the $CuO_{1-\delta}$ layer are represented in Fig.21, they correspond with δ-values equal to 0.33 and 0.67 respectively.

High resolution images were obtained which give direct evidence for the occurrence of the $3a_o$ structure (Fig.22a), as well as for $4a_o$ and $5a_o$ structures (Fig.22b,c). These images were found in beam heated specimens and therefore the corresponding compositions are uncertain. In view of the thermal treatment of the specimens their composition is presumably such that $\delta > 1/2$. Possible models are therefore as represented in Fig.21. The $3a_o$ diffraction pattern exhibits unsplit basic spots and the corresponding composition is therefore presumably $\delta > 1/2$ i.e. $\delta = 0.66$; i.e. the lower model of Fig.21b should apply.

5.4 The $2\sqrt{2}a_o \times 2\sqrt{2}a_o$ structure

The diffraction pattern of this type (Fig.23a) can be obtained in all specimens irrespective of their initial

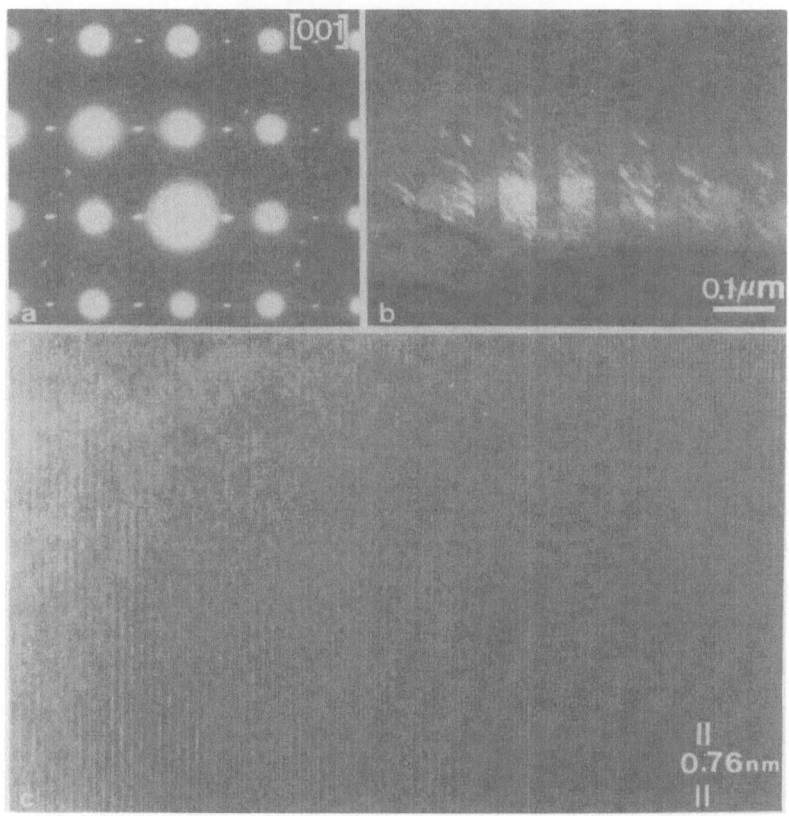

Fig.18 Constant stoichiometry cooled specimen with $\delta \sim 0.5$
(a) [001] zone diffraction pattern, note the sharp
superstructure spots; (b) dark field image made in a
superstructure spot: the bright patches are OII;
(c) high resolution image along the [001] zone

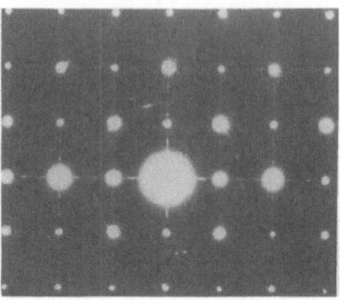

Fig.20 Electron diffraction pattern of a specimen exhibitir
two orientation variants of the 3a$_o$ structure. Note the
absence of slitting of the basic spots.

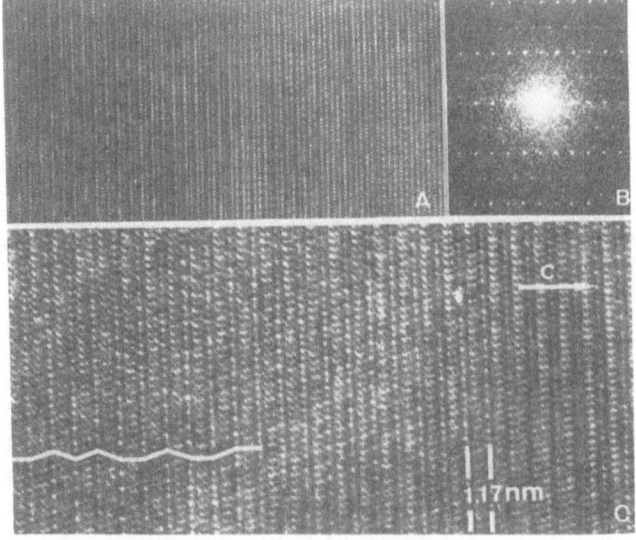

Fig.19 Alternative stacking in the OII structure of beam
treated specimen
(a) high resolution image at low magnification
(b) optical diffraction pattern of (a)
(c) high magnification of (a). Note the centering of the
    unit mesh.

composition by beam heating in the microscope to close
to the melting point. The diffraction pattern disappears
within a few minutes when maintaining the specimen at
microscope temperature, but it can easily be recovered
by a new beam heating-cooling cycle. This operation can

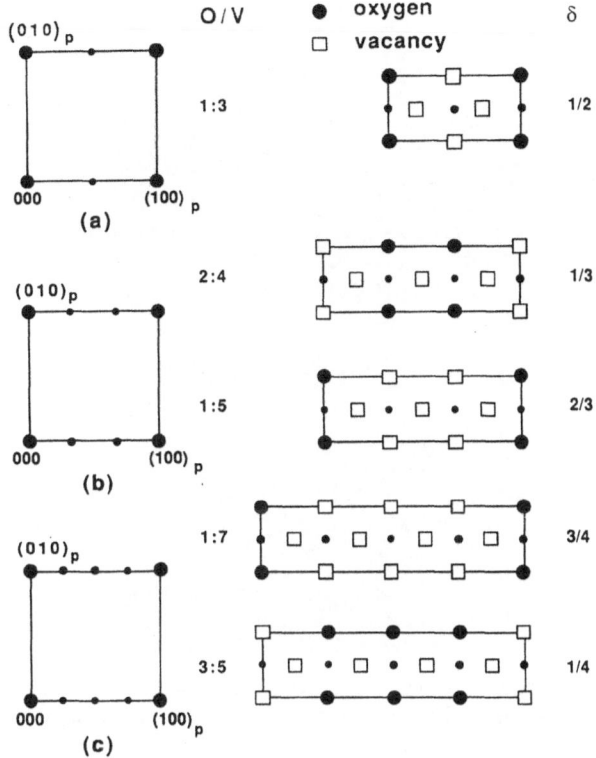

Fig.21 Idealized models for the $2a_o$, $3a_o$ and $4a_o$ structures and their corresponding diffraction patterns along [001]. For $3a_o$ and $4a_o$ two different models corresponding with two different δ values are compatible with the diffraction pattern.

be repeated a number of times. The rate of cooling is not important.
The superstructure spots are always sharp in the [001] zone, but they are due to continuous streaks along the [001]* direction. The intensity of the superstructure spots relative to that of the basic spots is much larger than for the na_o superstructures. No splitting of the basic spots is observed and no twins are detected, showing that the basic lattice is tetragonal although the structure models proposed in [9][10] are orthorhombic. The relative intensity of the superstructure spots with respect to basic spots far from the origin is larger than that for basic spots close to the origin; this strongly suggests that the superstructure is due to

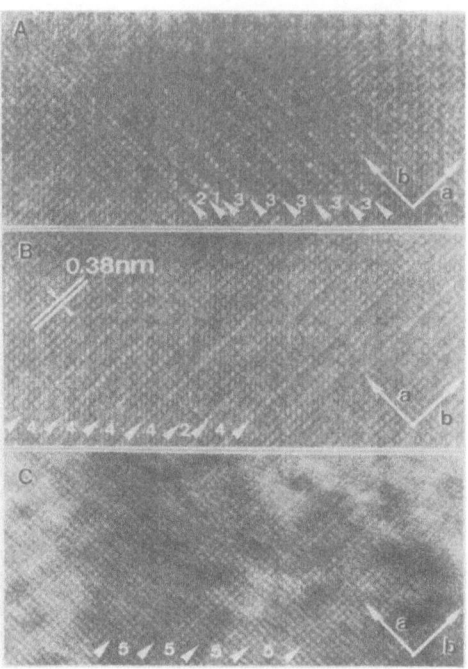

Fig.22 High resolution images of beam heated specimens
       giving evidence for the occurrence of na_o struc
       tures
       (a) $3a_o$ (b) $4a_o$ (c) $5a_o$

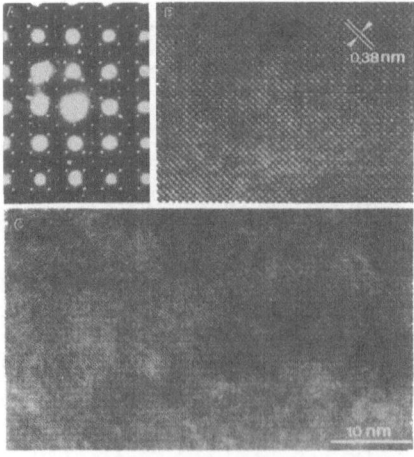

Fig.23 The $2\sqrt{2}a_o \times 2\sqrt{2}a_o$ structure
       (a) diffraction pattern along the [001] zone
       (b) low magnification high resolution image
       (c) high magnification high resolution image

periodic deformation rather than to ordering of oxygen atoms.

High resolution images of this structure are reproduced in Fig.23b,c.

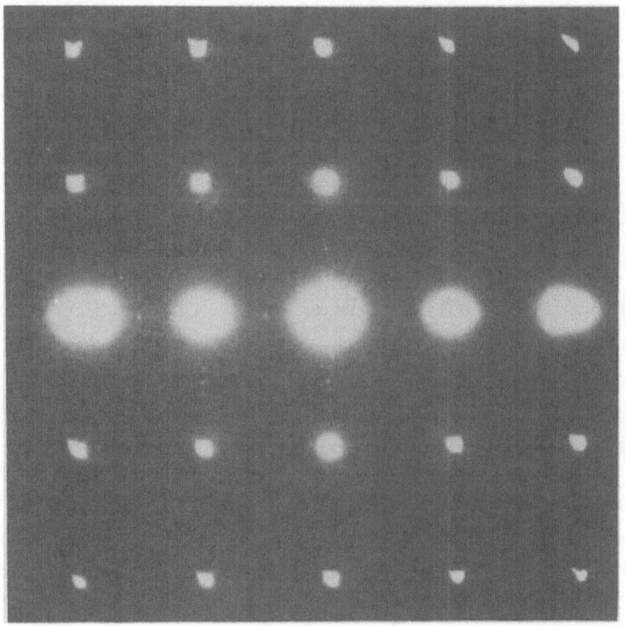

Fig. 24 Diffraction pattern along [001] of the $2a_o \times 2a_o$ structure.

5.5 The $2a_o \times 2a_o$ structure

In one case we obtained the diffraction pattern reproduced in Fig.24 in a specimen which was heat treated in the microscope. Although the zone axis was not exactly along·[001] the configuration of diffraction spots reveals unambigeously a $2a_o \times 2a_o$ unit mesh. Models which correspond with different δ values are represented in Fig.25.

6. DISCUSSION

Most of the observed diffraction patterns of super-structures are compatible with more than one composition. In Fig.21 we have represented schematically the different diffraction patterns and models with different δ-values which are compatible with the geometry and the symmetry of these diffraction patterns.

It is of course not possible to determine the arrangement of oxygen atoms and vacancies from the geometry of the diffraction pattern alone. However if the δ-value is known accurately the number of possible models is reduced. We can further restrict the number of plausible models using the well established fact that the basic lattice remains orthorhombic at room temperature for $0 < \delta < 0.65$ as judged from the splitting of the Bragg spots, which decreases with increasing value of δ, provided the specimen is ordered. Moreover, when carefully beam heating a specimen with $\delta \cong 0$ initially in the microscope it is possible to observe the diffraction patterns that appear in succession. Since under such conditions it is likely that oxygen is progressively released from the $CuO_{1-\delta}$ layer it is reasonable to assume that phases with increasing δ-values appear in succession. Combining all these considerations one can propose reasonable models for the phases.

Such models are of interest in trying to detect which structural features are significant for the phenomenon of superconductivity and which phases (or combination of phases) are responsible for the 60K superconducting plateau. The different arrangements contain for instance different densities of Cu-O-Cu-O chains along the b_o direction. If this feature is of importance one could expect to find a relationship between the superconductive properties and the presence of such chains. This is already suggested by the fact that the 90K transition is associated with the presence of a maximum density of these chains, whereas in the 60K phase, to be associated with the $2a_o$ (orthorhombic II) structure, the density of chains is only half of this maximum [11].

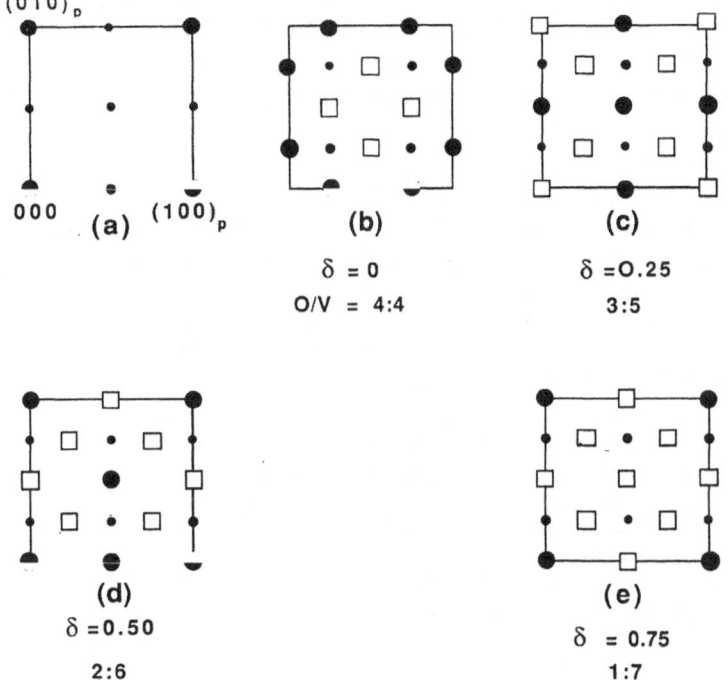

Fig.25 Possible models of the $2a_0 \times 2a_0$ structure corre
sponding with different δ values.

These considerations together with the observations
presented above allow to suggest a plausible succession
for the appearance (with increasing δ) of the different
structures. We shall now discuss the results of four
different types of specimen treatments.

6.1 Constant Stoichiometry (CS) specimens

The C.S. specimens with $\delta = 1/2$ exhibit prominently
the $2a_0$ structure in a well ordered form. Whereas in
previous work the $c_0{}^*$ rows of superstructure reflections
were always heavily streaked or were even continuous we
now find very sharp superstructure spots, leading to the
spatial structure model of Fig.17a. The $2a_0$ superstructure is the only one observed and it is detected in a
wide composition range from $\delta = 0.13$ to $\delta = 0.67$. For $\delta <$
0.13 we find exclusively the orthorhombic I structure
without any superstructure spots. For $\delta > 0.67$ we find

the tetragonal structure also without superstructure
spots.The average orthorhombicity decreases with increasing δ—value in the range $0 < \delta < 0.65$ because the
fraction of Cu-O-Cu-O chains along b_0 decreases with increasing δ.

In order to understand these results we assume that
for $\delta < 0.13$ and $\delta > 0.67$ the non-stoichiometry is ac-
commodated as randomly distributed linear clusters of
additional vacancies or ions oriented along the b_0 direction, of the type deduced from the diffuse scattering
[12]. The spacing between arrays is not regular enough
to produce pronounced superstructure spots. For $0 < \delta <$
1/2 the crystals apparently contain domains of the $2a_0$
structure and the orthorhombic I structure i.e. it consists in fact of two phases orthorhombic I and orthorhombic II, the volume fraction being such as to be
compatible with the macroscopic oxygen content. The dark
field images in the 1/2 0 0 spots provide evidence for
the occurrence of orthorhombic II in small patches, the
domains being largest in specimens with $\delta = 0.5$.

At the other end of the δ range $0.85 < \delta \leq 1$ the
crystal apparently contains a random distribution of the
remaining oxygen atoms in the $CuO_{1-\delta}$ layers, leading to
tetragonal symmetry. In the composition range $1/2 < \delta <$
1 the crystal again contains two phases the orthorhombic
II and the tetragonal phase. The size of the OII domains
however is much smaller than in material with $\delta = 0.5$.

6.2 Vacuum annealed specimens (V.A.)

The stepwise reduction of the specimen by heating in
vacuum, monitored by measuring the gas release, was used
to study the behaviour of Tc as a function of increasing
δ [13]. The Tc value was deduced from AC susceptibility
measurements. Such observations reveal the usual 90K
transition in nearly stochiometric samples and a 60K
transition in a broad range of reduced samples. By means
of electron diffraction we examined the phases present
in specimens having suffered different degrees of oxygen
loss.

As expected the 90K specimens only exhibit the orthorhombic I structure. The 60K specimens have prominently the $2a_0$ structure (but not as well ordered as in

the CS specimens with δ = 0.5) together with the OI structure. In specimens with a Tc intermediate between 60K and 90K we also found a poorly ordered $3a_O$ structure in the diffraction pattern. No other ordered superstructures were detected. As a consequence we believe that in these specimens the $3a_O$ structure has to be associated with a δ-value smaller than 1/2 and would thus be consistent with the δ = 1/3 model of Fig.21. This model consists in fact of rows of "normal", i.e. orthorhombic I unit cells alternating with rows of $2a_O$ superstructure.

6.3 Electron beam heated specimens (E.B.)

The largest variety of superstructures is found in specimens heat treated in the microscope vacuum by means of the electron beam.

The most prominent of the diffraction patterns obtained under these conditions is the $2\sqrt{2}a_O \times 2\sqrt{2}a_O$ pattern (Fig.23). It is always produced whatever be the initial specimen composition, provided the beam intensity or the number of pulses is large enough. It is therefore believed that this pattern corresponds with the largest δ-value which is compatible with the observed unit mesh i.e. the model of Fig.26b. For still larger δ-values the average distance between residual oxygen atoms presumably becomes too large to allow ordering. It should be noted that in C.S. specimens with such large δ-values no superstructures were found. The $2\sqrt{2}a_O \times 2\sqrt{2}a_O$ structure seems to be particular for E.B. specimens. It is possible that the ionizing effect of electron irradiation that accompanies the thermal effect is an essential feature for the formation of this superstructure.

The large relative intensities of the superstructure spots suggest that the oxygen sublattice does not make the main contribution to the intensities. It is well known that the perovskite structure is often periodically deformed. In the basic perovskite structure this can be attributed to the periodic tilting of corner linked octahedra. The situation is not so clear in the 1-2-3 compounds and needs further study by X-ray or neutron diffraction. However it is very likely that the large concentration of vacancies leads to a deformation pattern with a unit cell perhaps determined by the remaining oxygen atoms.Such tilt patterns are often uncor-

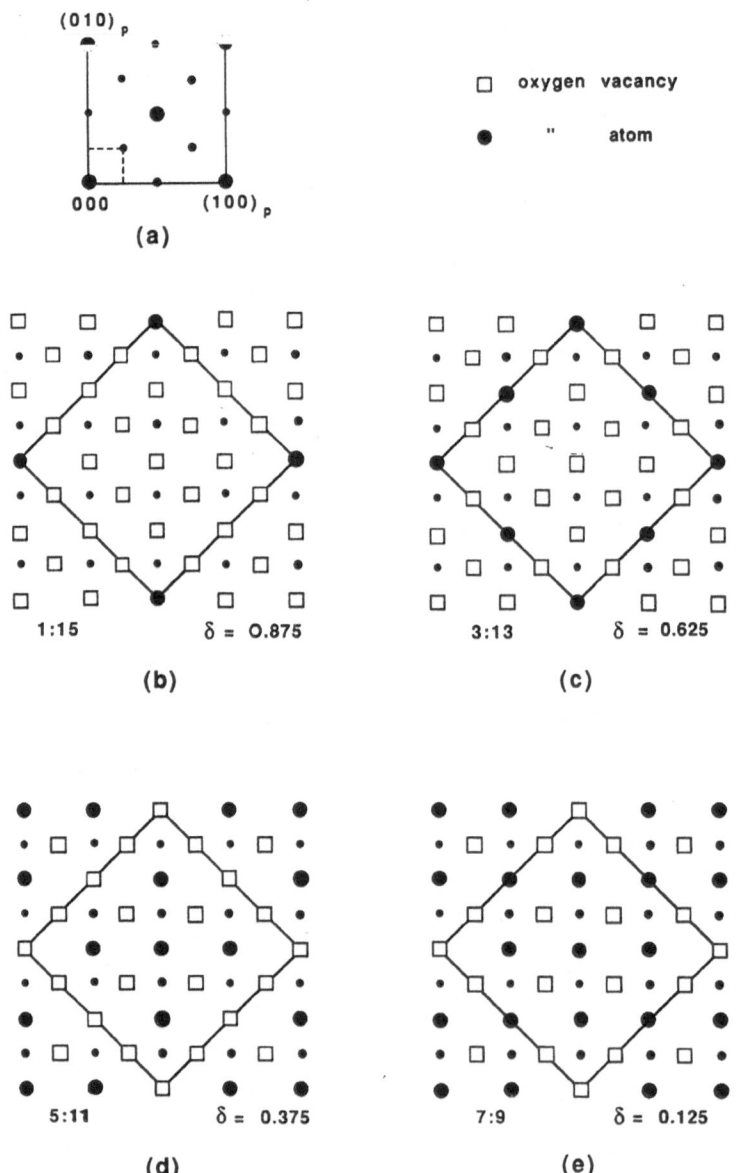

Fig.26 Different models compatible with the $2\sqrt{2}a_o \times 2\sqrt{2}a_o$ diffraction pattern. The lattice is apparently tetragonal whereas the models lead to orthorhombic structures. The corresponding δ values are indicated.

related in successive layers and could lead to the observed c_0^* streaking.

This suggestion is based on the observation that the superstructure reflections associated with the high order basic spots are relatively more intense with respect to the basic spots than those associated with low order spots; this is the signature for a deformation modulated superstructure [14]. In the $2a_0$ and the other one-dimensional superstructures the intensity of the superstructure spots decreases much more rapidly with increasing order.

In the high resolution images of the E.B. specimens treated to produce the $2\sqrt{2}a_0 \times 2\sqrt{2}a_0$ structure the $3a_0$, $4a_0$ and $5a_0$ structures were also found in small patches. It is believed that the na_0 patterns correspond in these specimens with the largest δ-values compatible with their unit meshes shown in Fig.21. We thus suggest in particular the occurrence of a second $3a_0$ structure, for $\delta > 1/2$ (see also 5.2). Also the alternative $2a_0$ structure with $c \cong 2c_0$ is found in such specimens.

Since the $2a_0 \times 2a_0$ pattern is produced in E.B. specimens just before the $2\sqrt{2}a_0 \times 2\sqrt{2}a_0$ pattern appears, it is believed that it corresponds with the model of Fig.24e of which the δ value is closest to that corresponding with the $2\sqrt{2}a_0 \times 2\sqrt{2}a_0$ structure. On continued beam heating the next stage after the appearance of the $2\sqrt{2}a_0 \times 2\sqrt{2}a_0$ structure consists in the formation of a simple cubic structure with a lattice parameter which is slightly smaller than that of $YBa_2Cu_3O_{7-\delta}$. This can be concluded from the composite diffraction pattern consisting of the superposition of the two patterns: the $2\sqrt{2}a_0 \times 2\sqrt{2}a_0$ pattern and the cubic pattern. Presumably the heavy metal sublattice becomes disordered and a defective cubic perovskite results in which Ba and Y are randomly distributed. This assumption would be consistent with the suggestion made earlier that at a temperature close to the melting point the heavy cations are disordered. This suggestion was based on the observation of domains with mutually perpendicular c-axis [15].

Summarizing : the E.B. specimens contain a number of superstructures. With increasing δ-values we find $2a_0$,

$3a_o$, $4a_o$, $5a_o$, $2a_o \times 2a_o$, $2\sqrt{2}a_o \times 2\sqrt{2}a_o$. Except for the $2a_o$ and the $2\sqrt{2}a_o \times 2\sqrt{2}a_o$ structures these superstructures occur in small patches only. In none of these na_o structures (n>2) the correlation between successive $CuO_{1-\delta}$ layers is sufficient to produce superstructure spots along the c_o^* direction.

From the fact that no structures other than OII occur in the C.S. specimens, but on the contrary occur frequently as more or less regularly spaced chain sequences in beam heated specimens we conclude that such structures must be metastable i.e. would be eliminated after a sufficiently long annealing treatment and only OII would be left.

The $3a_o$ configuration being the most abundantly observed one it is difficult to exclude at present that this might still be a stable phase in specimens with the adequate composition.

6.4 "In situ" heated specimens (I.S. specimens)

Experiments on C.S. materials were also performed "in situ" by heating in the microscope using the heating holder. The temperatures indicated by the thermocouple are presumably inaccurate, but their relative values are probably reliable. The following results may be meaningful. A specimen with $\delta = 0$ and exhibiting initially only the orthorhombic I structure is heated to ~400°C. The diffraction pattern of the same area is observed continuously at constant electron beam current. After a few minutes at 400°C weak superstructure spots due to two variants of the $2a_o$ structure become observable. Increasing the temperature to 500°C, causes these spots to dissappear again after a few minutes.

In another run the initial C.S. specimen was twinned and exhibited prominently the two variants of the $2a_o$ structure ($\delta = 0.5$). On increasing progressively the temperature it was found that the superstructure spots disappear first, whereas only at higher temperature the twinning spots and twin images disappear. At still higher temperatures the $2\sqrt{2}a_o \times 2\sqrt{2}a_o$ superstructure spots appear. This experiment suggests that the order-disorder transition occurs in several stages with slightly different characteristic temperatures. In the first stage, with increasing temperature, the correlation between the 2D arrangements in successive $CuO_{1-\delta}$ layers is lost; the

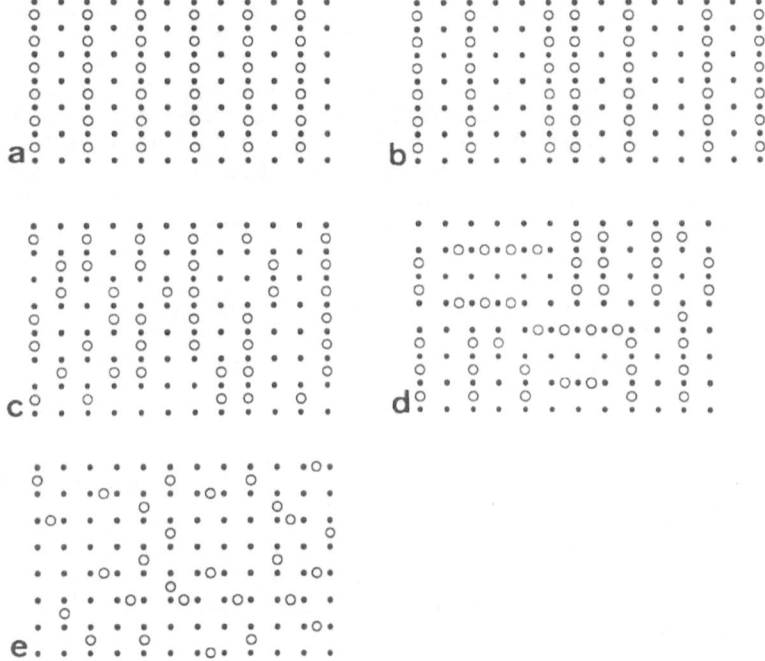

Fig.27 Schematic representation of the different stages
in the disordering of the oxygen sublattice on
increasing the temperature
(a) perfectly ordered $2a_o$ structure
(b) slightly disordered $2a_o$ structure; also a_o
and $3a_o$ spacings among chains occur occasionally
(c) the chains are broken up in finite segments,
but remain parallel
(d) the two chain directions become populated.
The specimen is macroscopically tetragonal, but
small orthorhombic domains remain
(e) the distribution of chain segments is
random,the structure is tetragonal.

h+1/2 0 1 rows of spots become streaked, but the 1/2 0 0
spots remain. In a subsequent stage the order among the
chains becomes short ranged; locally na_o patches are
formed which leads to broadening of the 1/2 0 0 spots.
The order within the chains, as well as their preferred
orientation along b_o is largely maintained, implying the
conservation of orthorhombicity (i.e. of twinning) al-
beit smaller than the original one. In a further stage
the chains are broken up and the two sets of oxygen
sites become occupied by chain segments, leading to a

diffraction pattern which is tetragonal on the average;
the diffraction spots are small crosses. These observations are consistent with the finding that in rapidly
cooled specimens the quenched-in tetragonal phase consists in fact of small orthorhombic domains, as a result of the formation of chain segments in two orientations [16]. True tetragonality is finally achieved when
all oxygen sites are occupied at random. This progressive disordering in stages is represented pictorially in
Fig.27.

The staged behaviour is consistent with a model proposed by de Fontaine et al. [17] in which the interaction V_2 between oxygen atoms mediated by copper i.e.
along the chains, is different from and stronger than
that which is not (V_3). Moreover these interactions have
a different sign. No interaction between successive
$CuO_{1-\delta}$ layers was considered; but it is undoubtedly much
weaker than the other interactions. The naïve assumption
that a given type of order will be destroyed the easier
the smaller the interaction energy that promotes this
type of order, is sufficient to understand the disordering in stages.

7. CONCLUSIONS

Our observations suggest that under the conditions
used, the most stable vacancy ordered superstructure is
the $2a_o$ (orthorhombic II) structure which occurs in two
stacking variants. Equilibrated specimens with a composition deviating from $\delta = 1/2$ contain in fact two
phases : orthorhombic I and II for $\delta < 1/2$ and orthorhombic II and tetragonal for $\delta > 1/2$. The 60K
plateau corresponding with $0.35 < \delta < 0.45$ being probably due to a mixture of OI and OII. The other superstructures are presumably only local and transient
structures.

The order-disorder transition occurs in steps; on increasing the temperature first the order among Cu-O-Cu-O
chains disappears and subsequently the order within
these chains is destroyed.

The $2\sqrt{2}a_o \times 2\sqrt{2}a_o$ pattern is possibly due to the combined effect of vacancy ordering and periodic deforma-
tion in the strongly oxygen deficient perovskite structure; this is suggested by the relatively large intensity of the superstructure spots.

ACKNOWLEDGEMENTS

This work has been performed in the framework of the Institute for Materials Science (IMS) of the University of Antwerp under an IUAP contract with the Ministry of Science Policy and with the financial support of the Belgian National Science Foundation (NFWO-IIKW). The authors would like to thank J. Reyes-Gasga, T. Krekels, H. Verwey, W.H.M. Bruggink for the use of illustrations from joint papers.

References

1. G. Van Tendeloo, H.W. Zandbergen, S. Amelinckx. Sol.Stat.Comm. **63**, 389 (1987).

    G. Van Tendeloo,H.W.Zandbergen, S. Amelinckx. Sol.Stat.Comm. **63**, 603 (1987).

2. J. Reyes-Gasga, T. Krekels,G. Van Tendeloo, J. Van Landuyt, S. Amelinckx, W.H.M. Bruggink, H. Verwey, Physica C, 1989 (in the press)

3. J. Reyes-Gasga, T. Krekels, G. Van Tendeloo, J. Van Landuyt, W.H.M. Bruggink, H. Verwey, S. Amelinckx, J.Sol.Stat.Comm., 1989 (in the press)

4. I.K. Schuller, D.G. Hinks, M.A. Beno, D.W. Capone, L. Soderholm, J.P. Locquet, Y. Bruynseraede, C.U. Segre, K. Zhang. Sol.Stat. Comm. **63**, 385 (1987).

5. M.A. Beno, L. Soderholm, D.W. Capone, D.G. Hinks, J.D. Jorgensen, J.K. Schuller, C.A. Seegre, K. Zhang, J.D. Grace. Appl.Phys.Lett. **51**, 57 (1987).

    R.J. Cava, B. Batlogg, R.B. Van Dover, D.W. Murphy, S. Sunshine, T. Siegrist, J.P. Remeira, E.A. Rietman, S. Zahuak, G.P. Espinosa. Phys.Rev.Lett. **58**, 1676 (1987).

    P.M. Grant, R.B. Beyers, E.M. Engler, G. Line, S.P. Parking, M.L. Ramirez, V.Y. Lee, A. Nazzal, J.E. Vazquez, R.J. Sanvay. Phys.Rev.**B35**, 7242 (1987).

R.M. Hazen, L.W. Finger, R.J. Angel, C.T. Prewitt, D.L. Ross, H.K. Mao, C.G. Hadidiacs, P.H. Mor, R.L. Meng, C.W. Shu. Phys. Rev. **B35**, 7238, (1987).

6. J. Sapriel, Phys.Rev. **B12**, 5128, 1975.

7. G. Van Tendeloo, S. Amelinckx, Video tape, University of Antwerp (Belgium).

8. H. Verweij, W.H.M. Bruggink. J.Phys.Chem.Sol.50, 75 (1989).

9. C. Chaillout, M.A. Alario-Franco, J.J. Capponi, J. Chenavas, J.L. Hodeau, M. Marezio. Phys.Rev.B **36** (1987).

    D.J. Werder, C.H. Chen, R.J. Cava, B. Batlogg. Phys.Rev.**B37** 2317 (1988).

10. M.A. Alario-Franco, C. Chaillout, J.J. Capponi, J. Chenavas.   Mat.Res.Bull. **22**, 1685 (1987).

11. R.J. Cava, B. Batlogg, C.H. Chen, E.A. Rietman, S.M. Zahurak, D. Werder, Phys.Rev.B **36**, 5719, 1987.

    R.J. Cava, B. Batlogg, C.H. Chen, E.A. Rietman, S.M. Zahurak, D. Werder, Nature, **329**, 423, 1989.

12. R. de Ridder, G. Van Tendeloo, D. Van Dyck, S. Amelinckx. Phys.Stat.Sol. (a) **38**, 663 (1976).

13. J. Van Acken et al. (submitted to Phys.Rev.).

14. J. Van Landuyt, G. Van Tendeloo, S. Amelinckx Phys.Stat.Sol. (a) **26**, K9 (1974).

15. H.W. Zandbergen, G. Van Tendeloo, T. Okabe, S. Amelinckx. Phys.Stat.Sol. (a) **103**, 45 (1987).

    G. Van Tendeloo, S. Amelinckx., Phys.Stat.Sol. (a) **103**, K_1-K_6, (1987).

16. G. Van Tendeloo, S. Amelinckx. J. Elect.Microsc. Tech. **8**, 285 (1988).

17. L.T. Wille, D. de Fontaine. Phys.Rev. **B37**, 2227 (1988).

    A. Berera, L.T. Wille, D. de Fontaine. Physica C, **153-155**, 598-601 (1988).

D. de Fontaine, M. Mann, G. Ceder
(preprint, to be published). Metastable states of
order in $Ya_2Cu_3O_{7-\delta}$.

D. de Fontaine, L.T. Wille, S.C. Moss
Phys.Rev.B, **36**, 5709 (1987).

L.T. Wille, A. Berera, D. de Fontaine
Phys.Rev.Letters **60**, 1065 (1988).

LOCAL ATOMIC DISPLACEMENTS IN HIGH T_c OXIDES

STUDIED BY PULSED NEUTRON SCATTERING

T. Egami, B. H. Toby, and W. Dmowski[*]

Department of Materials Science and Engineering and
Laboratory for Research on the Structure of Matter,
University of Pennsylvania, Philadelphia, PA 19104-
6272, USA

J. D. Jorgensen and D. G. Hinks

Materials Science Division, Argonne National
Laboratory, Argonne, IL 60439, USA

M. A. Subramanian, J. Gopalakrishnan, A. W. Sleight
and J. B. Parise

Central Research and Development, E. I. du Pont de
Nemours and Co., Wilmington, DE 19898, USA

INTRODUCTION

Various deviations from perfect periodicity in the atomic
structure are found in real superconducting oxides, due to extrinsic defects such as oxygen vacancies, site occupation disorder, or
twins. However, there are strong indications that these oxides
also contain significant <u>intrinsic</u> disorder which may play a role
in the superconductivity. A common indication of such disorder is
a large "temperature factor", or the Debye-Waller factor, in the
atomic structure which is often hardly dependent on temperature.
Such a large temperature factor often results not only from the
amplitudes of thermal vibration, but also from some static or
quasi-static atomic displacements which have not been properly
included in the structural analysis.

By using the method of atomic pair distribution function (PDF)
analysis of pulsed neutron powder diffraction data, we found in
some oxides these large thermal factors are in fact due to local
atomic displacements with only short range order. Unlike the more
conventional crystallographic methods of analysis, this method does
not presume the periodicity in the atomic structure, and does not

47

rely on the Bragg diffraction law. Instead, all the elastic
scattering intensities, including both the Bragg diffraction and
diffuse scattering intensities, are utilized in deriving the
distribution of the atomic distances, or the PDF [1]. Thus it is
an ideal method to study a structure with short range order. In
this paper we describe some of the results we obtained by this
method and discuss the effect of such an short range order on the
superconductivity.

EXPERIMENTAL METHOD

The method of PDF analysis has primarily been used in studying
the structure of amorphous materials for which the crystallographic
methods cannot be applied. We have found, however, that it is also
a very effective method to investigate the structure of crystalline
materials with internal disorder. In this method the normalized
scattering intensity, or the structure factor, $S(Q)$, where Q is the
magnitude of the scattering vector, or the momentum transfer during
the scattering, is Fourier transformed to obtain the PDF, $\rho(r)$, by

$$\rho(r) = \rho_0 + \frac{1}{2\pi^2 r} \int [S(Q) - 1]\sin(Qr)Q^2 dQ \qquad (1)$$

where ρ_0 is the average atomic density. The range of integration
should ideally be from 0 to ∞, but in practice it is limited to a
finite range. The accuracy of the PDF critically depends upon this
range of the integration, thus the availability of the scattering
data at high values of Q is important in increasing the accuracy.
The use of the pulsed neutron source is particularly beneficial for
this purpose, since the high intensity of epithermal neutrons
emanate from the pulsed source allow $S(Q)$ to be determined up to 30
or 40 Å^{-1}. The test of accuracy performed on crystalline Al showed
that the PDF thus determined is in agreement with the calculated
PDF over wide range of distances, at least up to 20 Å or more [2].
The experiments described in this paper were performed using the
SEPD and GPPD stpectrometers of the Intense Pulsed Neutron Source
(IPNS) of Argonne National Laboratory.

ATOMIC DISPLACEMENTS SUGGESTED BY THE PDF STUDIES

A. $Tl_2Ba_2CaCu_2O_8$

This compound is a n = 2 member of the $Tl_2Ba_2Ca_{n-1}Cu_nO_{4+2n}$
family [3] and has Tc of 105 K. The average structure has
tetragonal symmetry [4], however, the neutron PDF is not consistent
with this structure. In the tetragonal structure the distance
between thallium and oxygen atoms is 2.73 Å, which is larger than
the sum of the ionic radii, 2.28 Å [5] by 0.45 Å. Our PDF clearly
indicates that the Tl-O distance is not 2.73 Å, but has split into
short and long distances, with the short one, about 2.4 Å, being
much closer to the sum of the ionic radii [6]. The PDF is very
well explained by models in which both oxygen and thallium atoms
are displaced as shown in Fig. 1. In this structure the local

symmetry is lowered to either orthorhombic or monoclinic. The
correlations among the displacements are apparently only short
range, since they do not produce observable superlattice diffraction. These atomic displacements and concomitant lowering of local
symmetry can have significant effects on the electronic structure
of this compound. In this system charge carriers (holes) are
created in the CuO_2 plane most likely by the charge transfer to the
Tl-O band [7], which can be strongly affected by the position of Tl
and O. This compound represents a most convincing case of the
lowering of the local symmetry by displacive atomic short range
order. Unlike the substitutional short range order which is more
commonly found in many solids, the diffuse scattering due to
displacive order increases rather slowly with Q, becoming maximum
only at around $\pi/2\delta$, where δ is the magnitude of displacement.
Thus the PDF method is much more suited to detect the displacive
short range order than conventional methods.

Our recent study of the temperature dependence of the PDF
revealed an extremely intriguing anomaly at Tc [8]. As shown in
Figs. 2 and 3, the PDF, $\rho(r)$, shows pronounced changes between T =
100 K (below Tc) and T = 110 K (above Tc). The average of the
values of $\rho(r)$ at the peaks at 3.4 Å and 3.85 Å minus $\rho(r)$ at the
valley in-between (3.6 Å) shows a discontinuous jump at Tc, indicating some structural change is taking place at Tc. Since the
lattice constants determined by the Rietveld method vary smoothly
through Tc in agreement with the earlier study, this change is due
strictly to the internal short range atomic rearrangement which
does not change the average structure. The interpretation of this

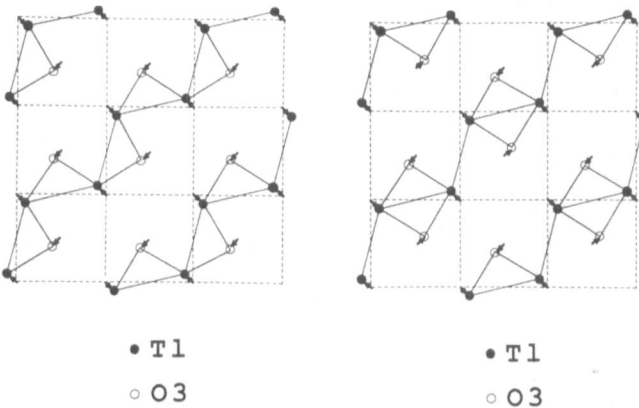

• Tl   • Tl

∘ O3   ∘ O3

Fig. 1. The displacements of Tl amd O atoms from the high-
symmetry sites in the Tl-O plane (corners and centers of the
square) for two idealized configurations. The ordering,
however, is only short range, and the real structure is most
likely the random mixture of microdomains of these two
configurations. (Ref. 6)

change in terms of a three- dimensional structure is still in
progress, and we are planning to repeat the measurement with even
better statistics. This represents the first clear evidence of
structural change at Tc. Other reports of such observation [9,10]
have been either indirect or irreproducible. It is thus quite
likely that the short range ordering has a direct link with the
superconducting phenomena, and further studies of the structural
change may lead to a better understanding of the mechanism of
superconductivity.

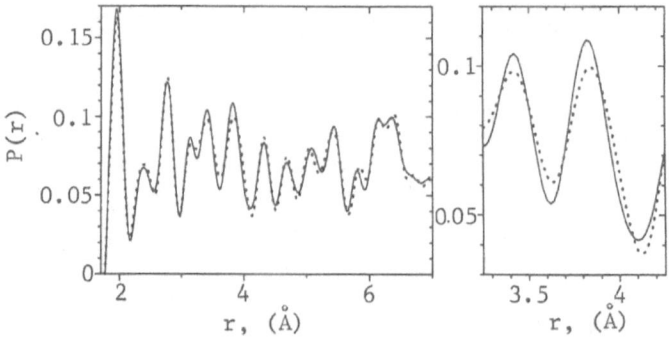

Fig. 2. The PDF of $Tl_2Ba_2CaCu_2O_8$ at T = 100 K (below Tc,
solid curve) and at T = 110 K (above Tc, dotted curve) [8].

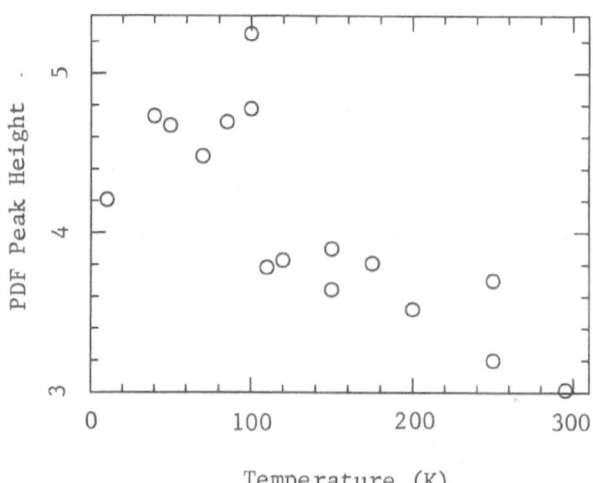

Fig. 3. The temperature dependence of the average peak
amplitude of the PDF peaks at 3.4 and 3.85 Å, indicating an
abrupt change at around Tc (105 K) [8].

B. Tl₂Ba₂CuO₆

B. $Tl_2Ba_2CuO_6$

This n = 1 member of the Tl compound family can be prepared in
either superconducting or semiconducting state depending upon the
synthesis condition [11,12]. Our preliminary study of this
compound in both states indicates that the superconducting solid,
reacted at 875°C, has much better short range order than the semiconducting one, reacted at 750°C, indicating that the superconductivity may be destroyed in one of the solids by the atomic level
disorder. It is possible that the difference in the electrical
behavior is due to a small difference (estimated to be less than
0.1 out of 6, if not zero) in the oxygen content, but the the
lattice constants of these two are virtually identical, suggesting
that they have almost the same composition. Thus the interpretation based upon randomness is favored. Since the samples we
studied are not completely single-phase our conclusion should be
considered preliminary. Further studies are being planned.

C. YBa₂Cu₃O₇₋δ

C. $YBa_2Cu_3O_{7-\delta}$

Unlike the other superconducting oxides we studied so far
including $La_{2-x}(Sr,Ba)_xCuO_4$ [13], this compound shows no clear sign
of displacive deviations from the average structure, although they
cannot be ruled out at a finer scale. Instead this system is known

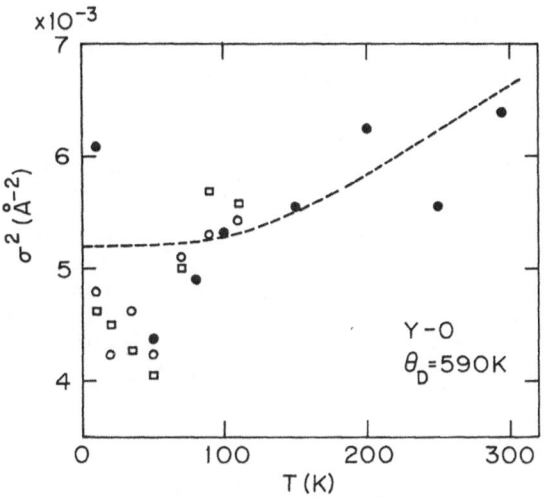

Fig. 4. The temperature dependence of the square of the
Gaussian peak width of the Y-O peak (2.4 Å) of $YBa_2Cu_3O_7$. The
dashed curve represents the Debye model with the Debye
temperature of 590 K. Open and closed symbols represent two
different data sets with different samples. The data deviate
from the Debye model below Tc (= 95 K).

to show vacancy ordering [14] with possible local orthorhombic
order even in the tetragonal phase ($\delta > 0.6$) [15]. Furthermore,
the height of the PDF peak corresponding to Y-O distance at 2.4 Å
shows anomalous behavior as shown in Fig. 4. Here σ is the
Gaussian width of the Y-O peak, and the dashed line represents the
Debye model with the Debye temperature of 590 K. The data points
significantly deviate from the Debye prediction below Tc (= 95 K).
The temperature dependence of other peaks, namely Cu-O and Ba-O
peaks are well accounted for by the Debye model. Just as the case
of Tl compounds in Fig. 3 the peak is higher below Tc implying more
structural correlation below Tc.

LOCAL STRUCTURAL INCOMPATIBILITY

The difference between the atomic distance required by the
structure and the ideal distance, which may be called the local
structural incompatibility [16], is an important origin of the
atomic displacements as we mentioned for the case of the 2-layer Tl
compound. In this case the lattice constant a is determined primarily by the Cu-O separation, which is strong and is almost in-
variant through all the cuprate superconductors, so that the Tl-O
distance in the high symmetry tetragonal structure is forced to be
longer than ideal.

It may be argued that the "ideal atomic distance" cannot be so
easily defined, thus the atomic incompatibility cannot be uniquely
defined. It is indeed only an approximation to define the ideal
distance as the sum of the ionic radii. However, the concept of
the atomic incompatibility can be put on a sound basis by defining
it in terms of the local atomic level stresses [16]. The atomic
level stresses are not zero even in perfect crystals as long as the
unit cell has more than one inequivalent atomic sites, and can be
calculated, at least in principle, from first principles [17].

This local atomic incompatibility explains the atomic displacements in Tl-cuprates, and possibly Bi-cuprates in which
similar displacements are suspected [18]. The reason why such
displacements remain only short range and do not develop into long
range order is unknown. One possibility is that the kinetics of
rearrangement slows down quickly with temperature, and the configuration becomes frozen in a partially disordered state. Such a
state occurs in mixed ionic crystals and molecular crystals, and is
known as the orientational glass state [19].

EFFECT OF RANDOMNESS

The atomic displacements resulting from the structural incompatibility can produce either the <u>long range</u> or <u>short range</u> structural effects. In the case of short range ordering it would be
important to differentiate the case when the structural correlation
length, λ, is <u>shorter</u> than the superconducting coherence length, ξ,
and the case when λ is <u>longer</u> than ξ.

A. Long range ordering

The structural distortions in La_2CuO_4 or $Ba_{1-x}K_xBiO_3$ and the incommensurate structural modulation in $Bi_2Sr_2CaCu_2O_8$ are likely to be the consequences of the structural incompatibility. In $(Ba,K)BiO_3$ the onset of the structural distortion coincides with the disappearance of superconductivity [20,21], while in Bi-Sr-Ca-Cu-O the long range modulation coexists with superconductivity.

B. Disorder ($\lambda < \xi$)

If λ is smaller than ξ, then the structural disorder may inhibit the superconducting phenomena. $Tl_2Ba_2CuO_6$ with Tc = 0 may be the case. Furthermore the reduction in Tc of $La_{2-x}Sr_xCuO_4$ when the content of Sr is more than 0.2 and the structure becomes tetragonal may also be due to the effect of disorder. In the tetragonal phase the apex oxygen of CuO_6 octahedron (02) is randomly displaced toward neighboring La [13], and this randomness can cause the supression of Tc. A rapid fall in Tc at around δ = 0.5 in $YBa_2Cu_3O_{7-\delta}$ may also be the effect of oxygen vacancy disorder in the chain (01), rather than the loss of the holes in the conducting plane.

C. Medium Range Order ($\lambda > \xi$)

If λ is larger than ξ, however, electrons do not see the disorder. This is probably the case for $Tl_2Ba_2CaCu_2O_8$. This situation is a most interesting one, since by limiting the ordering to short range it is possible to achieve a local order which may collapse when the order becomes long range. In other words this mechanism may keep a strong electron-phonon interaction from transforming the structure.

CONCLUSION

The atomic pair distribution analysis of the powder neutron scattering data has shown that the atomic structure of most of the high-Tc oxides is not strictly periodic, but there are substantial deviations from the periodicity due to local atomic displacements. These deviations are in large part simply due to the size mismatch effect, but may play an important role both in supressing the superconductivity by disorder, and in assisting it by creating environment in which strong electron-phonon interaction can be sustained. The anomaly in the PDF observed for Tl-Ba-Ca-Cu-O suggests that the short range order may have an active role in producing superconductivity. Further studies on temperature effects would clarify this point.

ACKNOWLEDGMENT

This work was supported by the National Science Foundation through DMR-8617950 and DMR-8519059 (U. of PA.) and U.S. Department of Energy, Basic Energy Science through W-31-109-ENG-38 (ANL). * Now at the Institute of Materials Science and Engineering,

Technical University of Warsaw, Warsaw, POLAND

REFERENCES

1. e.g. B. E. Warren, "X-Ray Diffraction" Addison-Wesley, Reading
    (1969).
2. S. Nanao, W. Dmowski, T. Egami, J. W. Richardson, Jr. and J. D.
    Jorgensen, Phys. Rev., B35, 435 (1987).
3. Z. Z. Sheng and A. M. Hermann, Nature, 332, 55 (1988).
4. M. A. Subramanian, J. C. Calabrese, C. C. Torardi, J.
    Gopalakrishnan, T. R. Askew, R. B. Flippen, K. J. Morrissey,
    U. Chowdhry and A. W. Sleight, Nature, 332, 420 (1988).
5. R. D. Shannon, Acta Cryst. A32, 751 (1976).
6. W. Dmowski, B. H. Toby, T. Egami, M. A. Subramanian, J.
    Gopalakrishnan and A. W. Sleight, Phys. Rev. Lett., 61, 2608
    (1988).
7. W. E. Pickett, Rev. Mod. Phys., 61, 433 (1989).
8. B. H. Toby, W. Dmowski, T. Egami, J. D. Jorgensen, M. A. Subra-
    manian, J. Gopalakrishnan and A. W. Sleight, unpublished.
9. P. M. Horn, D. T. Keane, G. A. Held, J. L. Jordan-Sweet, D. L.
    Kaiser, F. Holtzberg and T. M. Rice, Phys. Rev. Lett., 59,
    2772 (1987).
10. S. D. Conradson and I. D. Raistrick, Science, 243, 1340 (1989).
11. C. C. Torardi, M. A. Subramanian, J. C. Calabrese, J.
    Gopalakrishnan, E. M. McCarron, K. J. Morrisey, T. R. Askew,
    R. B. Flippen, U. Chowdhry and A. W. Sleight, Phys. Rev.,
    B38, 225 (1988).
12. A. W. Hewat, P. Bordet, J. J. Capponi, C. Chaillout, J.
    Chenavas, M. Godinho, E. A. Hewat, J. L. Hodeau and M.
    Marezio, Physica, C156, 369 (1988).
13. T. Egami, W. Dmowski, J. D. Jorgensen, D. G. Hinks, D. W.
    Capone, II, C. U. Segre and K. Zhang, Rev. Solid St. Sci.,
    1, 247 (1987) and in "High Temperature Superconductivity",
    eds. S. M. Bose and S. D. Tyagi, World Scientific, Singapore
    (1987) p. 101.
14. J. D. Jorgensen, H. Shaked, D. G. Hinks, B. Dabrowski, B. W.
    Veal, A. P. Paulikas, L. J. Nowicki, G. W. Crabtree, W. K.
    Kwok, L. H. Nunez and H. Claus, Physica C 153-155, 578
    (1988).
15. J. A. Gardner, H. T. Su, A. G. McKale, S. S. Kao, L. L. Peng,
    W. H. Warnes, J. A. Sommers, K. Athreya, H. Franzen and
    S.-J. Kim, Phys. Rev., B38, 11317 (1988).
16. T. Egami, J. Non-Cryst. Solids, 106, 207 (1988).
17. V. Vitek and T. Egami, phys. stat. sol., (b)144, 145 (1987).
18. C. C. Torardi, J. B. Parise, M. A. Subramanian, J.
    Gopalakrishnan and A. W. Sleight, Physica, C157, 115 (1989).
19. e.g. J. M. Roe, J. J. Rush, D. G. Hinks and S. Susman, Phys.
    Rev. Lett., 43, 1158 (1979).
20. R. J. Cava, B. Batlogg, J. J. Krajewski, R. C. Farrow, L. W.
    Rupp, Jr., A. E. White, K. T. Short, W. F. Peck, Jr. and T.
    Y. Kometani, Nature, 332, 814 (1988).
21. D. G. Hinks, B. Dabrowski, J. D. Jorgensen, A. W. Mitchell, D.
    R. Richards, Shiyou Pei and Donglu Shi, Nature, 333, 836
    (1988).

OXYGEN STOICHIOMETRY AND SUPERCONDUCTIVITY

IN CERAMICS

Y. Bruynseraede, J. Vanacken, B. Wuyts, J.-P. Locquet(°)
and Ivan K. Schuller*

Laboratorium voor Vaste Stof-Fysika en Magnetisme, Katholieke
Universiteit Leuven, 3030 Leuven, Belgium
* Physics Department - B019, University of California - San Diego
La Jolla, California 92093, USA

ABSTRACT

We have performed extensive gas evolution, X-ray diffraction and superconducting measurements on a series of high temperature ceramics as a function of oxygen composition. Constant rate gas evolution studies allow a unique identification of the oxygens evolving from the various inequivalent sites and permit a determination of their activation energies and frequency factors. In $YBa_2Cu_3O_{7-\delta}$ the evolution kinetics at fixed temperature and varying lengths of time is found to proceed by an initial large change in chain occupancy and a subsequent leveling off. The transition temperature determined from high frequency susceptibility measurements follows this behavior; a fast decrease in T_c up to a plateau around 60 K. The volume fraction of superconducting material decreases steadily, until the sample becomes non-superconducting.

INTRODUCTION

It is by now clearly established that changes in oxygen content can strongly affect the properties of high temperature ceramic superconductors. However, to date no clear cut picture has emerged which establishes uniquely the role of oxygen ordering, orthorhombicity, oxygen stoichiometry and their relationship to the superconducting properties.[1] The main reasons for these difficulties are that it is hard or perhaps impossible to determine in a unique way the presence of more than one oxygen phase. The establishment of oxygen ordering requires the application of complicated structural probes and in many cases the properties seem to depend delicately on a variety of preparation conditions.

(°) Present address : IBM Research Division, Säumerstrasse 4, 8803 Rüschlikon, Switzerland.

For the past two years we have embarked on a program dedicated to answering some of these questions. In this program, we have applied powder metallurgy and thin film techniques to prepare samples, x-ray and neutron diffraction together with gas evolution to establish the structure and stoichiometry and resistivity, magnetisation and high frequency susceptibility to study the superconducting properties.[2,3] These studies show that due to the complexity of the problem extensive systematic studies are needed in order to establish the general phenomenology of oxygen ordering in ceramics. In order to address these problems we have restricted our attention to $YBa_2Cu_3O_{7-\delta}$ and substitutions since this system is well understood structurally, much of its general phenomenology has been studied by a number of groups and it contains all interesting structural features, such as copper- oxygen chains and planes.

DETERMINATION OF OXYGEN STOICHIOMETRY

All detailed structural determinations done to date, especially the Rietveld refinements of neutron data, show consistently disagreements between calculation and experiment in certain ranges of k-vector.[4] The disagreements always occur in the same k-vector ranges and have the same character indicating that these are not due to subtle differences in the various samples but are possibly due to real structural inaccuracies, possibly due to a number of distinct oxygen phases. Unfortunately, neutron diffraction has not been able to establish the presence of various phases with different oxygen stoichiometry.[5]

We have applied a gas evolution technique, previously used to study hydrogen kinetics in amorphous silicon,[6] to study oxygen bonding in high temperature superconducting ceramics. These type of measurements allow a quantitative determination of oxygens in sites which have distinct binding energies. This measurement is complimentary to the traditional thermogravimetric measurement and it also allows the determination of oxygen binding energies.[7]

The sample (10 - 30 mg) is placed in an evacuated quartz tube. The temperature is then raised at a constant rate (typically $10°C/min$) and the pressure in the vessel is measured with a pressure gauge and analysed using a quadrupole mass spectrometer. The data are then corrected with the experimentally determined evolution from an empty tube. All corrections amount to less than 5 % . The number of evolved oxygen atoms is directly related to the time derivative of the pressure if a constant heating rate is applied.

Figure 1 shows a comparison of gas evolution from a number of high temperature superconductors. Note that the samples which contain Cu-O chains in their structure ($LaBa_2Cu_3O_{7+\delta}$, $YBa_2Cu_3O_{7-\delta}$, $Y_{0.8}Pr_{0.2}Ba_2Cu_3O_{7-\delta}$) exhibit an evolution peak centered around 600°C. A further comparison of this data with neutron scattering measurements clearly indicates that this peak corresponds to evolution from the Cu-O chains.[8] A binding energy of 1.2 eV is obtained from a fit assuming first order desorption, the broad width of the evolution peak arising possibly from the presence of various inequivalent sites. The high temperature peak cannot be uniquely identified, although it is thought that they arise from impurity phases and/or decomposition.

The oxygen desorption kinetics can be studied using this methodology very conveniently. In order to allow for proper ordering of the oxygens, low temperature (400°C), long time (0 - 100 hrs) anneals have been performed possibly while the sample is in the tetragonal phase.[9]

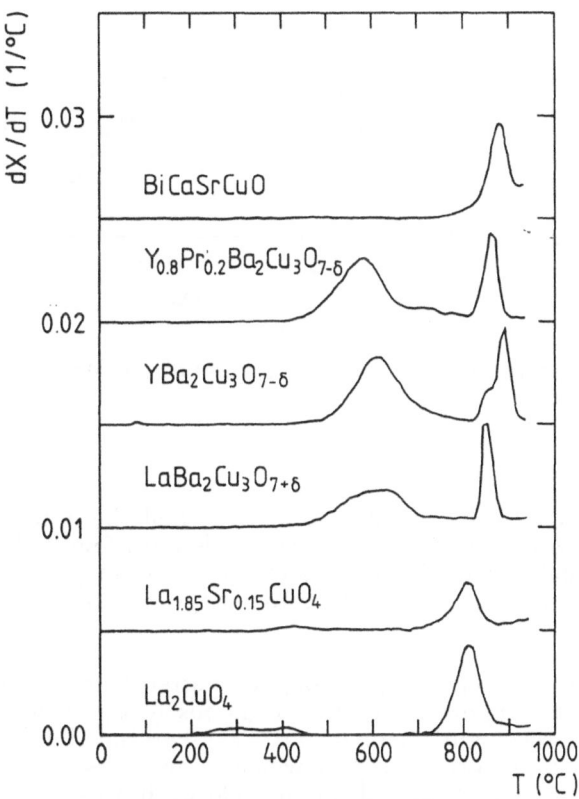

Fig. 1. Temperature derivative of the number of oxygen atoms evolving from different ceramic superconductors. (Curves are shifted upwards for clarity)

To avoid questions related to sample inhomogeneity one single sample was repeatedly heated for various lengths of times at 400°C in vacuum. After each annealing cycle the sample was slowly (100°C/hr) cooled to room temperature. Then the magnetic susceptibility was measured in cooling and heating from 300 K to 20 K. Once the superconducting properties were measured gas evolutions as described earlier were performed up to 950°C. Consequently, the

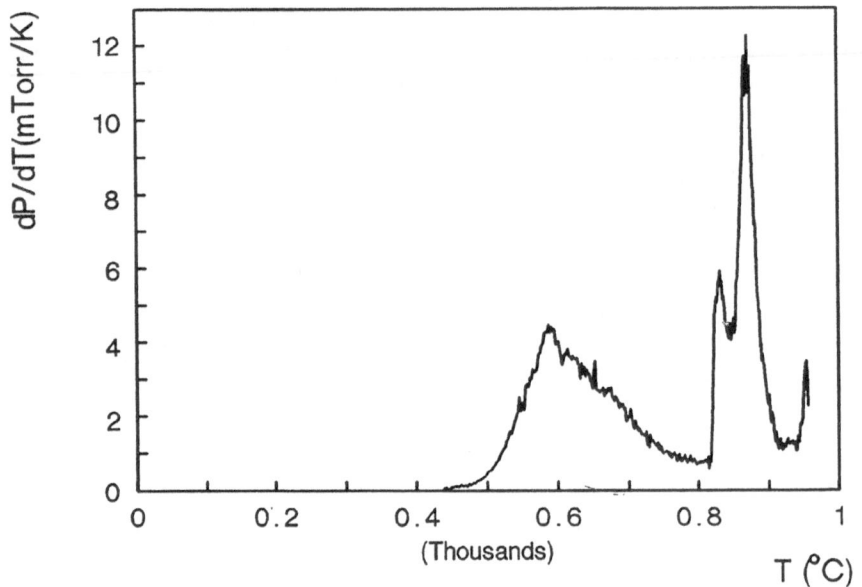

Fig. 2. Typical evolution spectrum for a sample that was annealed at 400°C for 3 hours.

sample was slow cooled to room temperature and the process was repeated again. It should be stressed at this stage that we found that the physical properties are quite critically dependent on the exact process, i.e. vacuum, anneal temperature etc., possibly due to changes in the structure during annealing.

Figure 2 shows a typical evolution curve after 3 hrs of anneal at 400°C. The first peak as explained earlier corresponds to the evolution from Cu-O chains, whereas the higher temperature peaks are possibly due to impurity phases. The most important feature which is connected to the superconductivity is the first peak, centered around 600°C. It was shown very early in the field of high temperature superconductors that the oxygen content and/or ordering of the Cu-O chains are correlated critically with the superconducting properties.[10,11]

Figure 3 shows the oxygen loss in $YBa_2Cu_3O_{7-\delta}$ under the lower (600°C) peak as a function of anneal time at 400°C. The evolution is characterized by a fast decrease in oxygen concentration in the first few hours of anneal followed by a plateau which is constant up to 100 hrs.

The difference in the oxygen kinetics at 400°C can be understood in light of the oxygen thermodynamics. The heat of solution of oxygen has a very strong dependence on the overall oxygen concentration. At high concentrations the heat of solution is small and therefore the oxygen is desorbed readily, at low concentrations the heat of solution is higher and therefore the desorption becomes very difficult in accordance with the results shown in Fig. 3.

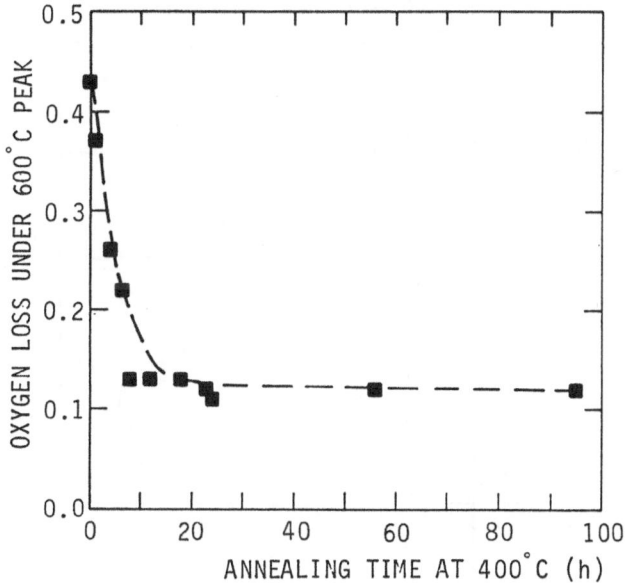

Fig. 3. Evolution of the O1 and O5 ("chains") occupancy versus annealing time at 400°C.

SUPERCONDUCTIVITY

The correlation of the superconducting properties with the gas evolution has been performed by measuring the high-frequency (32 MHz) susceptibility obtained from the frequency change of an RLC oscillator containing the sample. The superconducting transition is signaled by a sharp step in the susceptibility versus temperature curve. The normalized temperature derivative of this curve allows a precise determination of the onset, midpoint and completion of the superconducting transition. The area under the derivative curve is proportional to the volume fraction of the superconducting phase present.

Figure 4 shows a series of derivative curves for various annealing times. There is a systematic fast decrease of the transition temperature from 90 K to

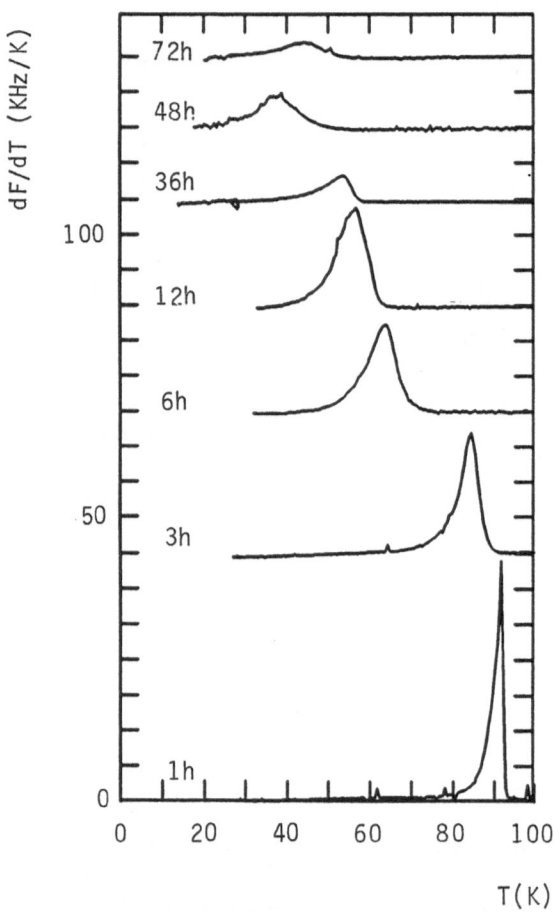

Fig. 4. Frequency change of the RLC oscillator circuit (proportional to the susceptibility change) versus temperature for samples with different annealing times at 400°C. (Curves are shifted upwards for clarity)

55 - 60 K and then Tc levels off (Fig. 5). Clearly this Tc behavior is correlated with the oxygen evolution curves (Fig. 3). The width of the transition from the normal to the superconducting state does not change considerably along the T_c versus annealing time curve (Fig. 4). On the Tc plateau, however, the volume fraction of the "60 K phase" decreases progressively for longer annealing times (Fig. 4). It should be pointed out that in some samples a two phase behavior was observed with the 90 K phase persisting for some annealing time. The details of this two phase behavior are presently under investigation.

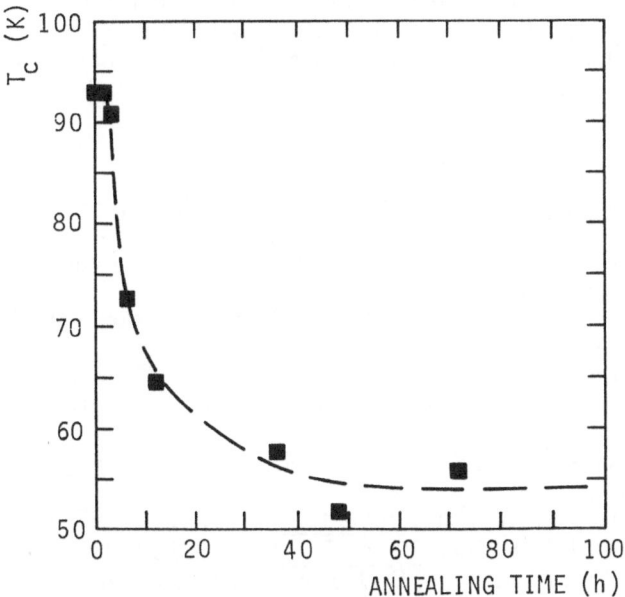

Fig. 5. Evolution of the transition temperature Tc versus annealing time at 400°C.

These results indicate that as the oxygen deficiency increases, a phase with an intermediate Tc between 90 and 60 K is formed. Further anneal of this phase, although it doesn't change the total oxygen stoichiometry, decreases the volume fraction of the phase. At some point percolation is lost and in a resistive measurement for instance the sample becomes normal.

In summary, we have shown that in $YBa_2Cu_3O_{7-\delta}$ long time anneals at 400°C produce samples with Tc's in the range 90 - 55 K. The kinetics proceeds by a fast decrease in oxygen concentration followed by a plateau which persists for anneals up to 100 hrs. The superconducting transition temperature follows this behavior, a fast decrease in Tc followed by a plateau around 55 - 60 K. Along the plateau however, the volume fraction of superconducting material decreases steadily, until the sample becomes nonsuperconducting.

ACKNOWLEDGEMENTS

This work is supported by the Belgian Inter University Institute for Nuclear Sciences (I.I.K.W.), the Inter University Attraction Poles (I.U.A.P.) and Concerted Action (G.O.A.) Programs (at K.U.L.) and the U.S. Office of Naval Research under contract number N00014-88K-0480 (at U.C.S.D.). International travel was provided by NATO. Two of us (J.V. and B.W.) would like to thank the Belgian I.W.O.N.L. and F.K.F.O. for financial support.

REFERENCES

1. For a review, see J.M. Tarascon and B.G. Bagley, Mater. Res. Bull. XIV, 53 (1989).
2. H. Strauven, J.-P. Locquet, O.B. Verbeke and Y. Bruynseraede, Solid State Commun. 65, 293 (1988).
3. J.-P. Locquet, J. Vanacken, B. Wuyts, Y. Bruynseraede and I.K. Schuller, Europhys. Lett. 7, 469 (1988).
4. see for instance M.A. Beno, L. Soderholm, D.W. Capone II, D.G. Hinks, J.D. Jorgensen, I.K. Schuller, C.U. Segre and J.D. Grace, Appl. Phys. Lett. 51, 57 (1987).
5. see for instance Ivan K. Schuller and J.D. Jorgensen, Mater. Res. Bull. XIV, 27 (1989).
6. E.C. Freeman and W. Paul, Phys. Rev. B 18, 4288 (1978).
7. S. Oguz and M.A. Paesler, Phys. Rev. B 22, 6213 (1980).
8. J.D. Jorgensen, M.A. Beno, D.G. Hinks, L. Soderholm, K.J. Volin, R.L. Hitterman, J.D. Grace, I.K. Schuller, C.U. Segre, K. Zhang and M.S. Kleefisch, Phys. Rev. B 36, 3608 (1987).
9. J.D. Jorgensen, H. Shaked, D.G. Hinks, B. Dabrowski, B.W. Veal, A.P. Paulikas, L.J. Nowicki, G.W. Crabtree, W.K. Kwok, L.H. Nunez and H. Claus, Physica C 153-155, 578 (1988).
10. I.K. Schuller, D.G. Hinks, M.A. Beno, D.W. Capone II, L. Soderholm, J.-P. Locquet, Y. Bruynseraede, C.U. Segre and K. Zhang, Solid State Commun. 63, 385 (1987).
11. G. Van Tendeloo, H.W. Zandbergen and S. Amelinckx, Solid State Commun. 63, 389 (1987).

OXYGEN ORDERING IN $YBa_2Cu_3O_{6+\delta}$:

A PHASE DIAGRAM CALCULATION

J.M. Sanchez* and J.L. Morán–López**

*Henry Krumb School of Mines
Columbia University
New York, N.Y. 10027

**Institute of Physics
Universidad Autónoma de San Luis Potosí
78000 San Luis Potosí, S.L.P., Mexico

ABSTRACT

The phase diagram of the ordered phases in $YBa_2Cu_3O_{6+\delta}$ as a function of oxygen concentration δ is calculated. The model takes as units the octahedra clusters formed by the O(1), O(4) and O(5) sites coordinated to the Cu(1) sites. A ground state analysis is carried out for δ in the range 0 to 1 assuming that the only structural units present are octahedra occupied by 0, 1 and 2 oxygen atoms in the basal plane, in addition to a repulsive Coulomb interaction between O atoms not mediated by Cu(1) atoms. The model accounts for several of the experimentally observed ordered structures. At finite temperatures, the model Hamiltonian is solved within the octahedron approximation of the cluster variation method. The effect of the O(4) vacancies on the equilibrium phase diagram is analysed.

INTRODUCTION

After the intensive study of recent years, the main structural properties of the high temperature superconductors as a function of oxygen content are now well established. For example, the orthorhombic phase of $YBa_2Cu_3O_{6+\delta}$, shown in Fig. 1, consists of CuO_2 planes, formed by the atoms labeled Cu(2), O(2) and O(3), and of CuO_δ basal planes, formed by Cu(1) and O(1). In this phase five (two) differet oxygen (copper) sites can be distinguished. On the other hand, the number of diffrent oxygen sites in the tetragonal phase reduces to three, with the O(1) and O(5) sites and the O(2) and O(3) sites becoming equivalent.

The strong dependence of the superconducting critical temperature on oxygen content in these ceramics was recognized and experimentally confirmed very early by several groups. A detailed neutron diffraction study [1] showed that, at constant partial pressure, the oxygen content varies continuosly with temperature due primarily to the removal or addition of oxygen to the O(1) and O(5), and to a lesser extent, O(4) sites. The experimental evidence suggests that the amount of oxygen and its ordering in the basal plane play a key role in the superconducting properties of these materials. In particular, it appears that the superconducting state is achieved *only* in the orthorhombic phase with oxygen concentrations approximately in the range $0.5 \le \delta \le 1$.

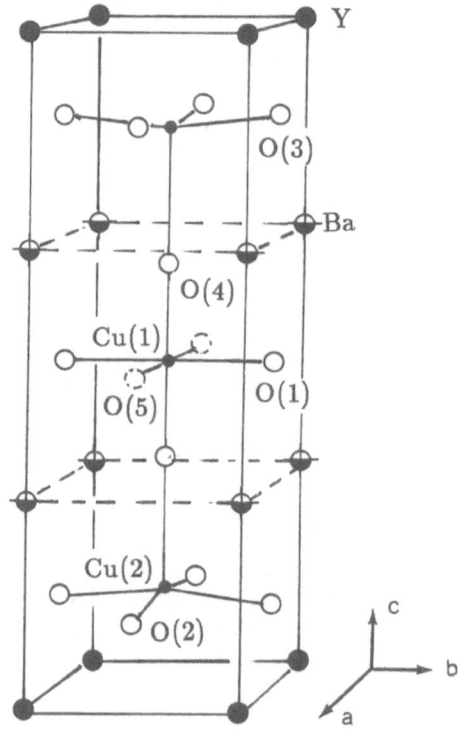

Fig. 1. The crystal structure of $YBa_2 Cu_3 O_{6+\delta}$.

The occurence of several ordered phase has also been experimentally observed [2-5] for different values of the oxygen stoichiometry δ. The highest value of Tc is observed for high symmetry ordered phase with $\delta=1$. In this phase, all the O(1) sites are occupied and all the O(5) sites are vacant, thus forming a pattern of oxygen chains (•) alternatating with vacancy chains (○). Electron [2] and X-ray diffraction [3] experiments revealed the presence of a lower symmetry ordered phase for oxygen stoichiometry $\delta = 0.5$. In that phase, the oxygen atoms are also ordered along chains separated by three chains of vacancies ($\bullet\circ\circ\circ\bullet\circ\circ\circ\bullet$). More recently, complex ordered structures with relatively large unit cells have been reported at stoichiometries $\delta \simeq 1/8$ and $7/8$ [4], and with reciprocal space periodicity $q=1/3$ and $2/5$ [6,7].

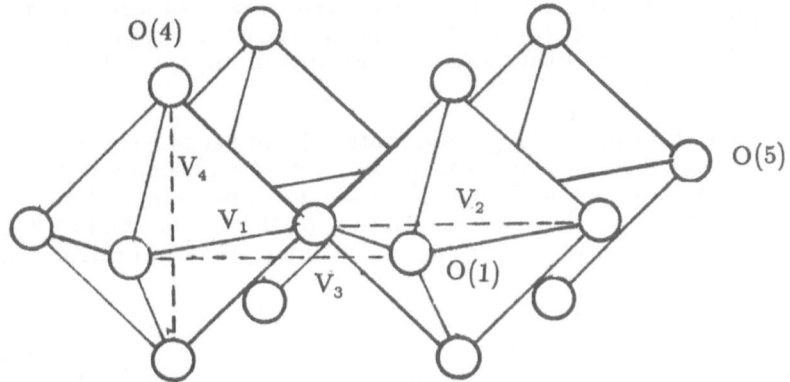

Fig. 2. The pseudo-three dimensional lattice made up of octahedra.

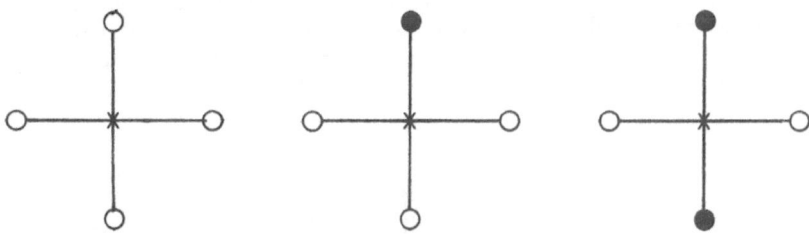

Fig. 3. Schematic representation of the octahedral units with 0, 1 and 2 oxygen atoms (closed circles) in the basal plane.

In order to explain some of the phases observed, Kachaturyan and Morris [8] proposed a model based on the *maximum amplitude principle*. This principle, which has been used to explain the appearance of ordered compounds in transition metal oxides, is satisfied when atoms and their vacancies form layered structures composed of alternating filled and vacant planes with periodic faults on every $(2n+1)$ plane. Applied to the $YBa_2Cu_3O_{6+\delta}$, this model predicts the occurence of ordered structures for $\delta=1-n/(2n+1)$. The ordered patterns for $\delta=1$ and $\delta=1/2$ are those discussed above and for $\delta=2/3$ and $3/5$

are ● ○ ● ○ ○ ○ ● ○ ● ○ ○ ○ and ● ○ ● ○ ○ ○ ● ○ ○ ○ ● ○ ●, respectively. This model does not predict, however, the observed ordered structures for $\delta=1/8$ and 7/8 [4].

The ordered phases with $\delta=0.5$ and 1, called Ortho–II and I respectively, have been studied using in a two dimensional Ising model with anisotropic second neighbor interactions [9]. Due to the short range of interactions, this model cannot account for the more complex ordered structures observed in $YBa_2 Cu_3 O_{6+\delta}$. Furthermore, the the experimentally observed O(4) vacancies, and their effect on the realtionship between the oxygen stoichiometry and the tetragonal–to–orthorhombic phase transition, has not been included in this theoretical study.

Here we investigate the ground states of the pseudo–three dimensional system formed by the octahedra mentioned above and shown in Fig. 2. Furthermore, using the octahedron as the basic unit, we carry out a finite temperature study using the cluster variation method [10]. In particular, we characterize the effect of O(4) vacancies on the equilibrium phase diagram.

GROUND STATE

In order to investigate the ground states in the range $0< \delta <1$, we consider the possible arrangements of octahedral units with both O(4) sites occupied and with 0, 1 or 2 oxygens on the basal plane sites. The three resulting units, projected on the basal plane, are shown in Fig. 3 and they will be referred as units 0, 1 and 2. These units exclude the occurence nearest–neighbor (nn) O atoms and, therefore, a strong nn repulsive interaction between oxygen atoms is implicitly assumed in the model. Allowing for all possible orientations of these units, there are seven different configurations, or *octahedral species*, to be arranged on a square (or tetragonal) two dimensional lattice (see Fig. 4). Furthermore, we assume that the relevant range of interactions

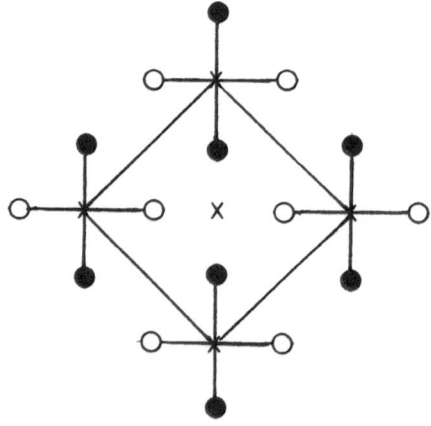

Fig. 4. Array of octahedra of the type 2 in a square lattice.

does not exceed that of square unit cell formed by four non–overlaping octahedra at each corner (see Fig. 4; note that the central octahedron in the unit
cell is determined by those at unit cell corners). The ground state problem is
reduced to that of a seven component square lattice in which the interactions
between them are anisotropic.

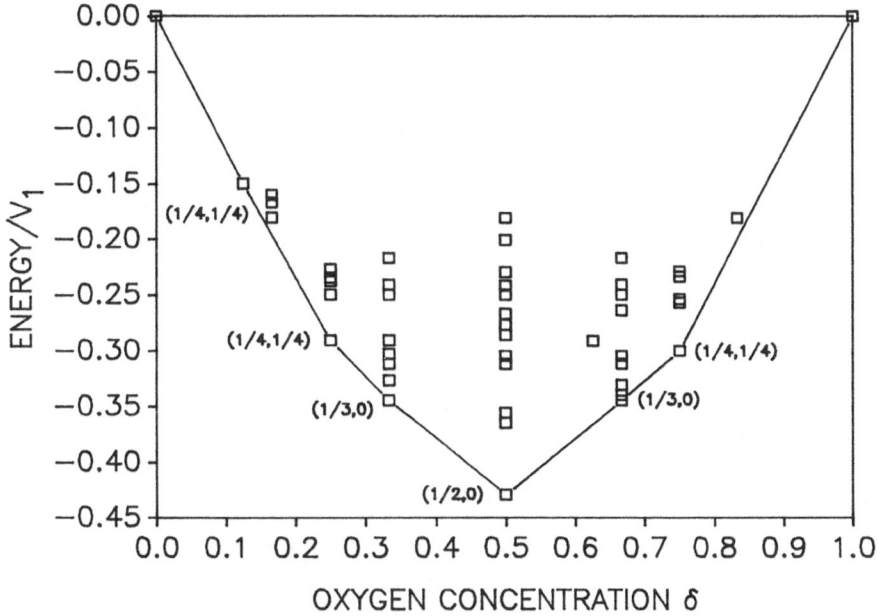

Fig. 5. Energy vs. oxygen concentration for the various ordered configurations. The solid line joins the ground states for $\Delta = 0.1$,
and $V_5'/V_5 = 0.5$.

In order to complete the ground state analysis, a specific form for the
intra– and inter–octahedron interactions is needed. The relevant intra–octahedron energy is given by $\Delta = E_0 - 2E_1 + E_2$, where E_i is the energy of the
type-i octahedron. We assume that the inter–octahedron interactions can be
represented well by an unscreened Coulomb potential between oxygen ions on
different octahedra. Furthermore, two types of fifth neighbor O–O pairs are
distinguished depending on whether they have (V_5) or not (V_5') a Cu atom
in between. A set of 51 possible structures was obtained by enumerating the
distinct arrangements of 4×4 octahedral units.

A plot of the energy versus δ, assuming $\Delta = 0.1$ and $V_5'/V_5 = 0.5$, is shown
in Fig. 5 for all the structures. The position of the Bragg peaks in the reciprocal space are indicated for the ground states (connected by a straight line)

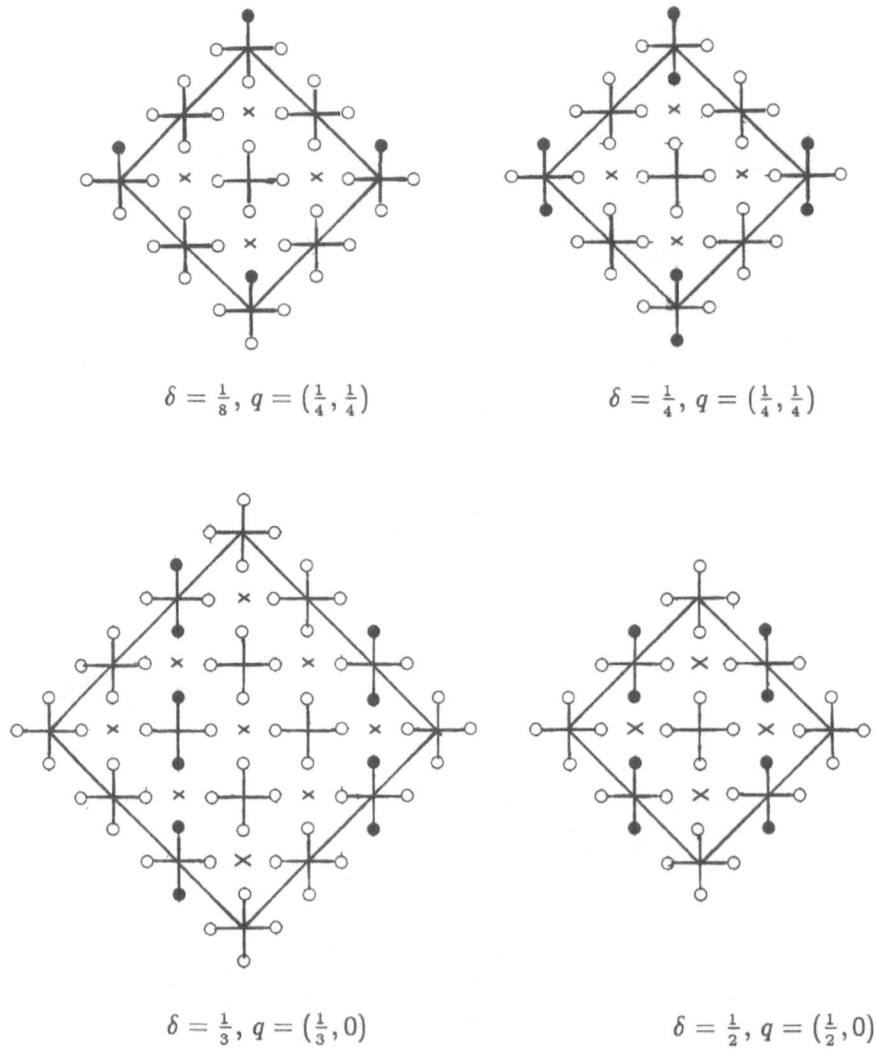

$\delta = \frac{1}{8}, q = \left(\frac{1}{4}, \frac{1}{4}\right)$ $\delta = \frac{1}{4}, q = \left(\frac{1}{4}, \frac{1}{4}\right)$

$\delta = \frac{1}{3}, q = \left(\frac{1}{3}, 0\right)$ $\delta = \frac{1}{2}, q = \left(\frac{1}{2}, 0\right)$

Fig. 6. Unit cells of the ground states corresponding to Fig. 5, for $\delta = \frac{1}{8}$, $\frac{1}{4}$, $\frac{1}{3}$ and $\frac{1}{2}$. The unit cells for $\delta = \frac{2}{3}$ and $\frac{3}{4}$ can be obtained from those of $\delta = \frac{1}{3}$ and $\frac{1}{4}$ by exchanging the 2 by 0–units.

which occur for $\delta=1/8$, 1/4, 1/3, 1/2, 2/3, and 3/4. Their respective unit cells are shown in Fig. 6. The structures for $\delta=1/8$, 1/3, and 1/2 have been observed experimentally [4-6]. Finally, we point out that the q=2/5 structure is not predicted by the model and, presumably, will require interactions beyond the range allowed here in order to become stable.

THE EFFECT OF THE O(4) VACANCIES ON THE PHASE DIAGRAM

It is now well documented that the $YBa_2Cu_3O_{6+\delta}$ system undergoes an orthorhombic to tetragonal phase transition at high temperatures, driven by the ordering of oxygen atoms in the basal plane. The high temperature tetragonal phase is characterized by equal oxygen occupancy in the O(1) and O(5) sites. Initially, it was belived that changes in oxygen site occupancy involved exclusively the O(1) and O(5) sites [1]. However, recent neutron diffraction experiments [11] have revealed that the oxygen atoms at the O(4) sites, which lie above and bellow the basal plane, become partially vacant and may influence the phase transition through their effect on overall oxygen stoichiometry.

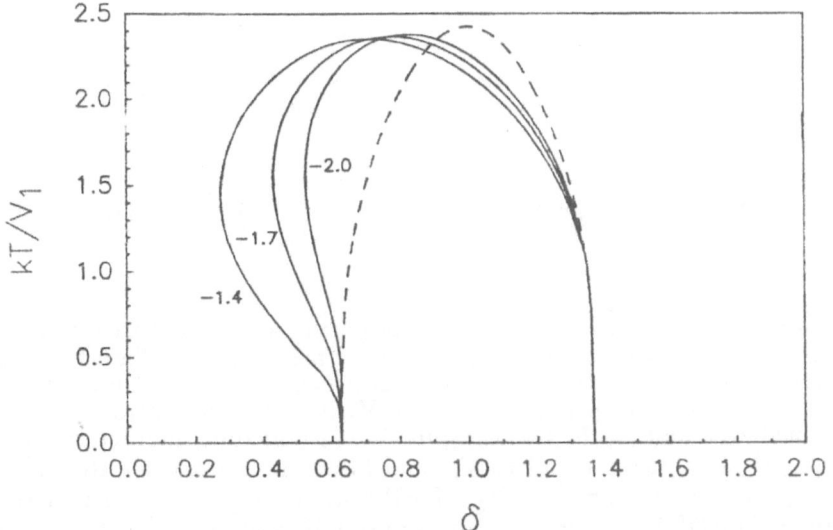

Fig. 7. Phase diagram kT/V_1 vs. δ, calculated including the O(4) vacancies for $\Delta\epsilon$=-1.4, -1.7 and -2. The dashed line corresponds to the case where $\Delta\epsilon \rightarrow \infty$.

In a previous publication [12] we studied the oxygen order–disorder transition in the basal plane using a two dimensional Ising model with only nn repulsive interactions. Within this model, the calculated phase diagram only displays the ortho–I and tetragonal (disordered) phases. Despite the drastic simplifying assumptions of the model, it was possible to account reasonably well for the experimental data on oxygen site occupancies at high temperatures.

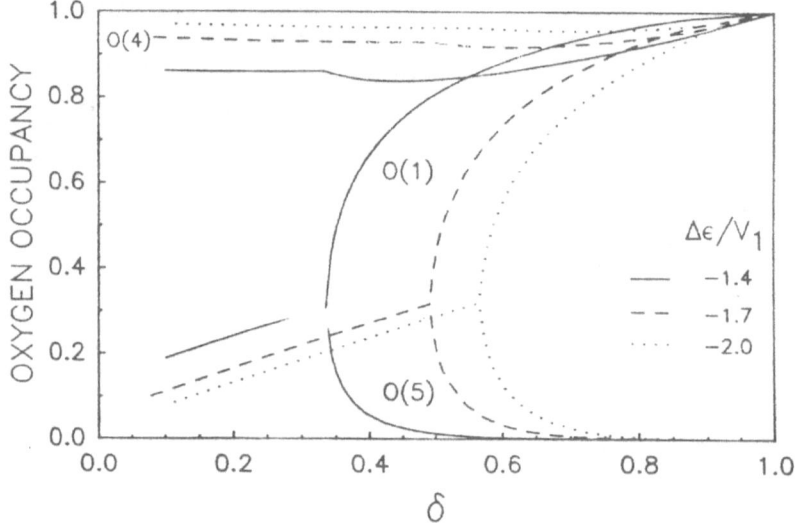

Fig. 8. Oxygen occupancy as a function of δ at the O(1), O(4) and O(5) sites at $kT/V_1 = 2.0$ and $\Delta\epsilon = -1.4, -1.7$, and -2.

Here, we extend the model by investigating a pseudo–three dimensional system made of basic octahedron units shown in Fig. 2. The oxygen interactions in the basal plane are pairwise between nearest–neighbors (V_1) and between next–nearest–neighbors V_2 and V_3 for, respectively, oxygen–oxygen pairs mediated and not by Cu(1) atom. We also include interactions between O(4)–O(1) sites, taken equal to V_1, and between two oxygen atoms located at O(4) sites, taken equal to V_2. Furthermore, in order to model the site occupancy of O(4) sites, we introduce in our Hamiltonian a difference in the on–site potential $\Delta\epsilon = \epsilon_{O(4)} - \epsilon_{O(1)}$ between O(4) and O(1) (or O(5)) sites.

The calculated phase diagrams are shown in Fig. 7 for the case of a repulsive nearest–neighbor interaction V_1 only. The dashed curves corresponds to the two dimensional square (rectangular) lattice results where the effect of the O(4) vacancies is ignored. The full line corresponds to the pseudo–three dimensional case, which includes the effect of O(4) vacancies on the phase

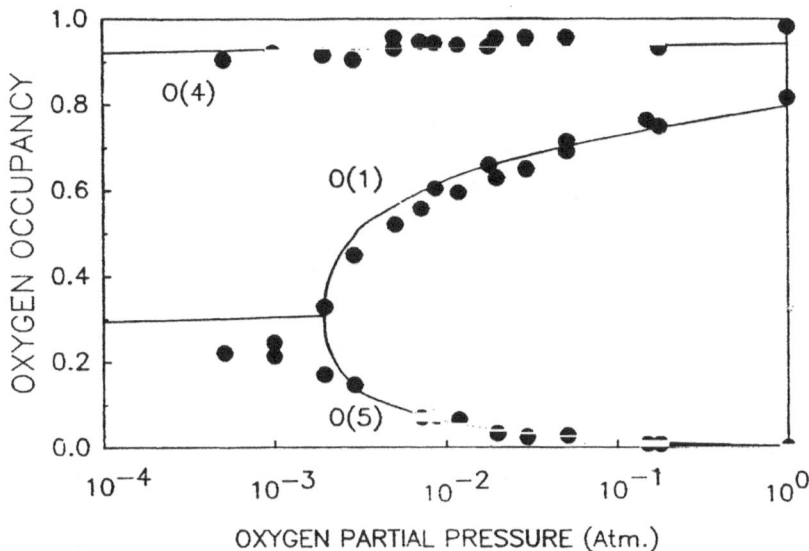

Fig. 9. Oxygen occupancy as a function of partial pressure, calculated (solid line) and experimentally determined [11] (solid circles).

transition, for different values of the on–site potential $\Delta\epsilon$. Large negative values of $\Delta\epsilon$ favor the $O(4)$ occupancy. The main effect on the phase diagram is to produce a shift of the second order transition lines to lower values of δ. This shift may explain the fact that the transition is experimentally seen to occur at a value of δ approximately equal to 0.4 [1,11], rather than the value of 0.6 suggested by the percolation limit of the square lattice.

The oxygen occupancy at the basal plane sites $O(1)$ and $O(5)$ and the off basal plane $O(4)$ sites is shown in Fig. 8. The interaction parameters taken here are those of figure 1 and the temperature was chosen at $kT/V_1=2.0$. As mentioned above large values of $\Delta\epsilon$ do not allow the formation of vacancies at the off basal plane sites. Here one can clearly see that as more $O(4)$ vacancies are formed, the larger is the shift of the transition temperature to lower values of δ.

The calculated and experimentally determined oxygen occupancy at the $O(1)$, $O(4)$ and $O(5)$ sites, as a function of oxygen partial pressure is shown in Figure 9. The overall agreement between the model and experiment is very good. The only portion of the experimental results that can not be well reproduced is the one at low partial pressures. The parameters used to fit the experiments are $V_1=1$, $V_2=-0.25$, $V_3=V_4=0.25$ and $\Delta\epsilon=-1.2$. The upper part of the phase diagram corresponding to the parameters used to fit the experimental data is shown in Figure 10. The low temperature portion of

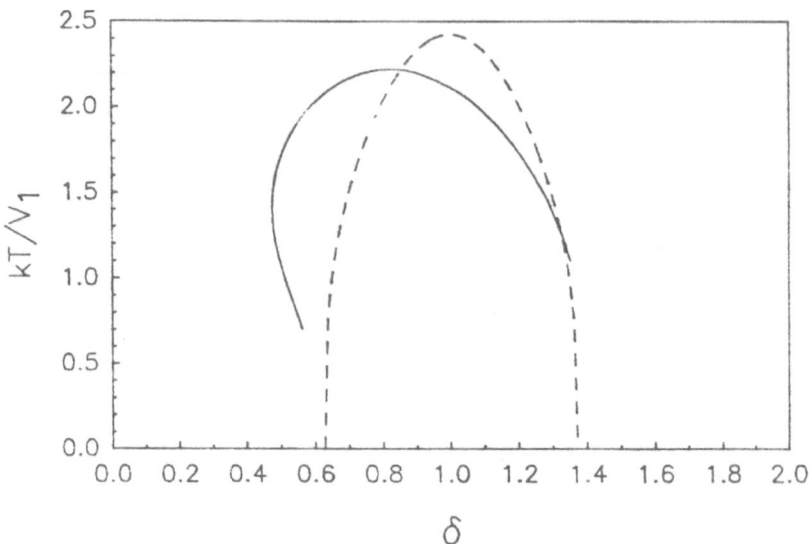

Fig. 10. High temperature portion of the order–disorder phase diagram (solid line) calculated with the parameters used to fit the experimental data of Fig. 9. The phase diagram obtained when O(4) vacancies are ignored is also shown (dashed line) for comparison.

the phase diagram is not shown since the ground state is the Ortho–II and requires additional calculations.

ACKNOWLEDGEMENTS: This work was supported in part by the National Science Foundation through Grants No. DMR–85–10594 and INT–88–14613, by DGICSA–SEP (Mexico) through Grant No. C89–08–0081, and by PRDCT of the Organization of American States.

REFERENCES

1. J.D. Jorgensen, M.A. Beno, D.G. Hinks, L. Soderholm, K.J. Volin, R.L. Hitterman, J.D. Grace, I.K. Schuller, C.U. Segre, K. Zhang, and M.S. Kleefisch, *Phys. Rev. B* **36**, 3608 (1987).
2. C. Chaillout, M.A. Alario-Franco, J.J. Capponi, J. Chenevas, and M. Marezio, *Solid State Commun.*, **65**, 283 (1988).
3. R.M. Fleming, L.F. Schneemeyer, P.K. Gallagher, B. Batlogg, L.P. Rupp, and J.V. Waszczak, *Phys. Rev. B* **37**, 7920 (1988).
4. M.A. Alario-Franco, J.J. Capponi, C. Chaillout, J. Chenevas, and M. Marezio, *Proc. Materials Research Society Meeting*, Boston 1987.
5. G. van Tanderloo, H.W. Zandbergen, and S. Amelinckx, *Solid State Commun.*, **63**, 289 (1987).

6. C.H. Chen, D.J. Werder, L.F. Scheeemeyer, P.K. Gallagher, and J.V. Waszczak, *Phys. Rev. B* **38**, 2888 (1988).
7. D.J. Werder, C.H. Chen, R.J. Cava, and B. Batlogg, *Phys. Rev. B* **38**, 5130 (1988).
8. A.G. Khachaturyan and J.W. Morris, Jr., *Phys. Rev. Lett.*, **61**, 215 (1988).
9. L.T. Wille, A. Berera, and D. de Fontaine, *Phys. Rev. Lett.*, **60**, 1065 (1988).
10. R. Kikuchi, *Phys. Rev.*, **81**, 988 (1951).
11. J.D. Jorgensen, H. Shaked, D.G. Hinks, B. Dabrowski, B.W¿ Veal, A.P. Paulikas, L.J. Nowicki, G.W. Crabtree, W.K. Kwok, L.H. Nunez and H. Claus, *Proceedings of the International Conference on High Temperature Superconductors: Materials and Mechanisms of Superconductivity*, Interlaken, Switzerland, February 1988.
12. J.M. Sanchez, F. Mejía–Lira, and J.L. Morán–López, *Phys. Rev. B* **37**, 3678 (1988).

STATES OF OXYGEN ORDERING IN YBa$_2$Cu$_3$O$_z$

D. de Fontaine

Department of Materials Science
and Mineral Engineering
University of California
Berkeley, CA 94720

INTRODUCTION

The compound YBa$_2$Cu$_3$O$_z$ (1-2-3) was the first one whose superconducting transition temperature exceeded the technologically important temperature of 77K[1]. Actually, the 1-2-3 compound at close to oxygen stoichiometry z=7 became superconducting at about T$_c$=90K, with T$_c$ dropping by increments as z decreased from the value 7.

In this compound, the transition temperature is closely related to the oxygen content but is also thought to be influenced by the arrangement of occupied and vacant oxygen sites in the CuO$_2$ mirror plane of the structure, i.e. the plane located between Ba ions. In the 90K superconducting phase, occupied sites form O-Cu-O chains, and the resulting three-dimensional structure has orthorhombic symmetry. When available sites in the mirror plane are occupied statistically by O ions, the resulting structure has tetragonal symmetry and the material is non-superconducting. It thus appears that the tetragonal to orthorhombic transition in the 1-2-3 compound can be modeled by an order-disorder transition in the mirror plane. It is the object of this paper to review the properties of this transition and to derive appropriate phase diagrams pertaining to oxygen ordering in the plane. Much pertinent information concerning oxygen ordering in 1-2-3 is reviewed in a recent paper by Beyers and Shaw[2].

THE MODEL

It follows from the foregoing that attention must be focussed on the mirror plane, a schematic representation of which is shown in Fig. 1[3]. Open circles denote oxygen sites which may be vacant or occupied; black dots refer to Cu ion positions, assumed to be always occupied. The oxygen sites form two

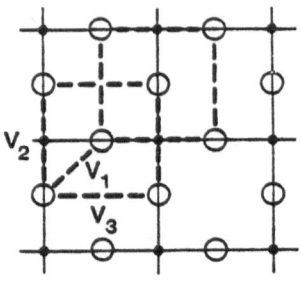

- Cu
- ① 04 : α Sublattice
- ⊖ 05 : β Sublattice

Fig. 1. Model for perovskite mirror plane with effective pair interactions V_n indicated.

interpenetrating square lattices and, if the Cu ions did not break the symmetry, the filled/empty oxygen network could be treated as a classical two-dimensional Ising model.

To simplify the thermodynamics, only three effective pair interactions (EPI) will be considered: V_1 which couples the two sublattices, V_2 which is a second neighbor interaction mediated by Cu, and V_3 which is the other next neighbor interaction. These three EPI's are expected to be the dominant ones.

EPI's are defined formally by

$$V_n = \frac{1}{4}(W_{OO}^{(n)} + W_{\square\square}^{(n)} - 2W_{O\square}^{(n)}) \qquad (n = 1,2,3)$$

where, for given pair spacing (n), $W_{IJ}^{(n)}$ represents the total energy of an otherwise disordered solution of O (oxygen) and $\square$ (vacancies) on the plane but containing the designated IJ pair at the specified sites. These energies can be calculated in principle by performing electronic band structure calculations. Such calculations have not yet been performed but plausibility arguments[4] indicate that V_1 must be positive, V_2 negative, of comparable magnitude, and V_3 negative but of lesser magnitude. From Eq. (1), it is seen that positive interactions favor unlike pairs, and negative interactions like pairs.

Several remarks concerning these EPI's are in order. Firstly, the V_n are *effective* interactions and must not be confused with *pair potentials*. Equation (1) clearly indicates that the EPI are calculated by taking differences of total energies. Actually, although pair potentials converge slowly with pair spacing, if at all, EPI's converge rapidly in metals[5], as if the long-range interactions canceled out by taking differences. Secondly, although the thermodynamic model proposed is a two-dimensional one, the full three-dimensional character of the system is introduced in the V_n interactions. Finally, it is important to allow second neighbor asymmetry: $V_2 \neq V_3$. Other proposed models which assume $V_2 = V_3$ are clearly deficient since they neglect the presence of Cu in the mirror plane, clearly a serious omission.

GROUND STATE ANALYSIS

Regardless of the values of the V_n's, it is instructive to perform a stability analysis of a completely disordered solid solution of oxygen and vacant sites. The second order term of a Landau expansion of the system's free energy in terms of amplitudes of oxygen "concentration waves" must have extrema at "special positions" in k-space which, in the present case, are points $<0,0>$, $<1/2,0>$ and $<1/2,1/2>$.[3] Which of these points present an actual minimum depends on the ratios $x = V_2/V_1$ and $y = V_3/V_1$. Results of the stability analysis are shown in Fig. 2: in each sector of the (x,y) plane is indicated which special point produces a free energy minimum. In particular, the $<0,0>$ special point indicates that an oxygen/vacancy wave of infinite wavelength modulates one of the sublattices and another, 180° out of phase with the first (because of $V_1 > 0$) modulates the other sublattice. In other words, one sublattice is filled by oxygen atoms and the other remains empty. The familiar chain structure of the orthorhombic phase results.

Actual ground states of order can also be investigated for all possible values of the ratios x, y. The results are indicated in Fig. 3[6]. For this range of EPI's, ordered structures are found only at stoichiometries $c_0 = 1/2$ or $1/4$ (or $3/4$), where c_0 is the oxygen atom fraction in the mirror plane. In Fig. 3a ($c_0 = 1/2$), the familiar chain structure is depicted in the $<0,0>$ instability region. More complicated structures are found in other regions but for x,y ratios which are not relevant to the present system according to the qualitative discussion given above. At $c_0 = 1/4$, complete phase separation is predicted for (x,y) in the fourth quadrant (Fig. 3b), and two new cell doubling structures (p2mm) are found in the second and third quadrants. Actually, the structure with "b" O-Cu-O and □-Cu-□ chain alternating along "a" has been observed experimentally in $YBa_2Cu_3O_z$ for $z \cong 6.5$[7-13], which corresponds to $c_0 = 1/4$ if it is assumed that all oxygen depletion occurs within the mirror plane. Clearly, if the EPI's are weakly dependent on concentration, the conditions for stability at low temperatures for both the single-chain structure, now called Ortho I, and the alternating chain structure, now called Ortho II, are $V_1 > 0$, $V_2 < 0$ and $0 < V_3 < \frac{1}{2} V_1$.

PHASE DIAGRAMS

The basic premise on which the present model rests is that the true three-dimensional order-disorder (oxygen-vacancy) phase transformations which occur in $a_2Cu_3O_z$ can be mapped onto a two-dimensional Ising model which is isotropic in the first-neighbor interactions and anisotropic in the second-neighbor interactions. The two-dimensional Ising model on a square lattice with first-neighbor interaction in zero field can be solved exactly but, when higher interactions are introduced, or when the applied field (magnetic, chemical potential difference) is non-zero, no closed form solution exists so that approximate methods must be used. Square lattice phase diagrams for first and second neighbor interactions have been

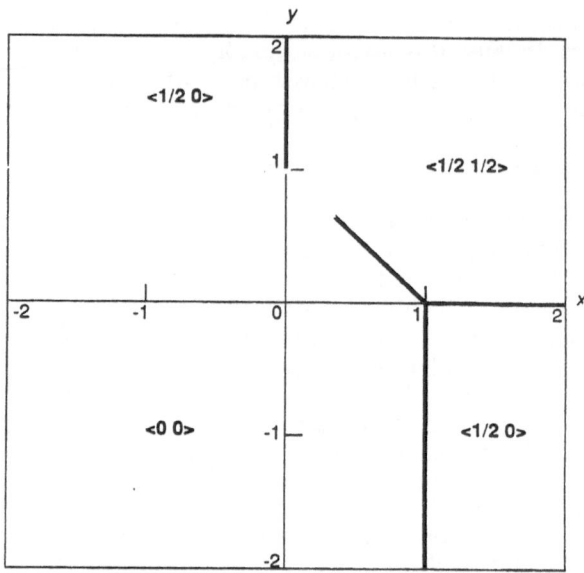

Fig. 2.   Ordering instability map for $V_1>0$ (ordering first-neighbor interaction
Coordinates are the ratios $x=V_2/V_1$, $y=V_3/V_1$. Open circles indicate
parameter ratios for which phase diagrams were calculated previously by
other methods. Closed circles (numbered) indicate parameters chosen for
present calculations.

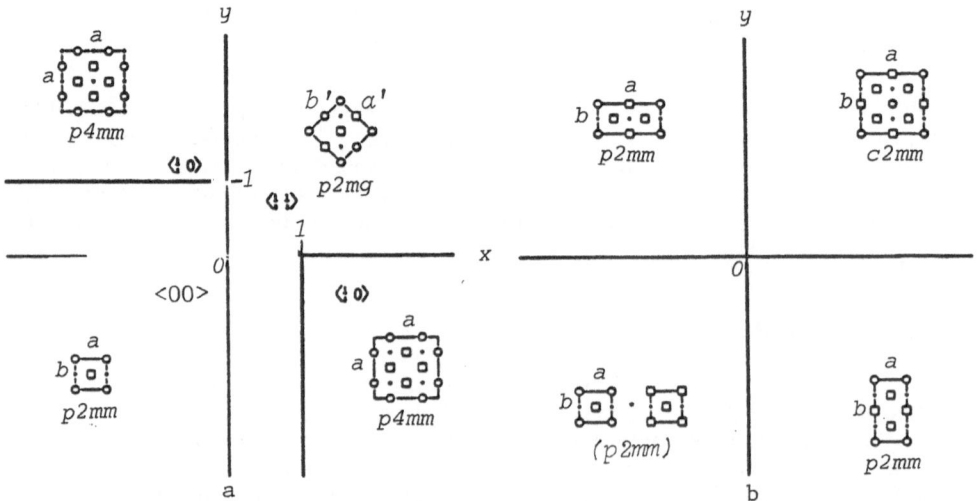

Fig. 3.   Ground states as a function of $x=V_2/V_1$, $y=V_3/V_1$, at oxygen
concentration: (a) 0.50 (b) 0.25.

calculated previously by renormalization group[14] and Monte Carlo[15] techniques. Recently[16,17], it was found that the suggested assymmetric model belonged to the same universality class as the Ashkin-Teller model[18] for which certain global phase diagrams have been derived[19]. This latter model considers a square lattice whose sites may be occupied by four types of objects in equal abundance with two distinct EPI's: one between like and one between unlike neighbors.

For the present calculations, the cluster variation method (CVM) of Kikuchi[20] was adopted. First, a test case was computed, that corresponding to $V_2=V_3=-\frac{1}{2}V_1$ ($V_1>0$) since previous phase diagram calculations were available[14,15] for these "symmetric" values of parameters. Results of the calculation have been reported earlier[21] and details of the CVM applied to this case were described elsewhere[22]. The representative point characterizing this phase diagram in parameter space is shown in Fig. 2 as "point #1" by a dot (our calculations) inside an open circle (previous calculations pertaining to symmetric cases).

The CVM phase diagram agreed closely with renormalization group[14] and Monte Carlo[15] results: a line of second order transitions separates the disordered (Tetragonal, planar point group 4mm) from the ordered (Ortho I, planar point group mm) phases and this line terminates at a tricritical point below which complete phase separation occurs. For the renormalization group and Monte Carlo calculations, the second-order transition line tends to be steeper just above the tricritical point and the top of the miscibility gap is also flatter than it is for the CVM calculations. The latter discrepancy is due to the fact that the CVM is, itself, a mean field theory, though a considerably improved one, and it tends to give classical exponents at transitions. It is known, however, that tricritical exponents are highly non-classical in two-dimensions.

CVM calculations performed for parameter ratios at point #2 on the stability map (Fig. 2), away from the diagonal x=y, produce a phase diagram which is very similar to that described above, and requires no further discussion. When V_3 is made to change from negative to small positive values, however, qualitative changes occur: a cell doubling phase, Ortho II, appears near stoichiometry $c_o=1/4$, as predicted by ground states analysis. Two phase diagrams have been computed with representative point #3 and #4 on the stability map.

The phase diagrams corresponding to point #4 is shown in Fig.4 [24]. In this figure, the phase boundary lines were terminated when the CVM free energy minimization failed to converge numerically. The most prominent feature of this diagram is the Ortho II phase region pertaining to a new, stable, equilibrium ordered phase, for which there is now ample experimental evidence.

Experimentally determined Tetra.↔ Ortho. I transition points have also been plotted on the phase diagram of Figs. 4 (closed circles). These points were constructed as follows: the data, obtained by a group from the Oak Ridge National Laboratory[25] give Tetra.↔ Ortho. I transition temperatures as a function of oxygen concentration at five different oxygen partial pressures. The data point for $p_{o_2}=0.2$

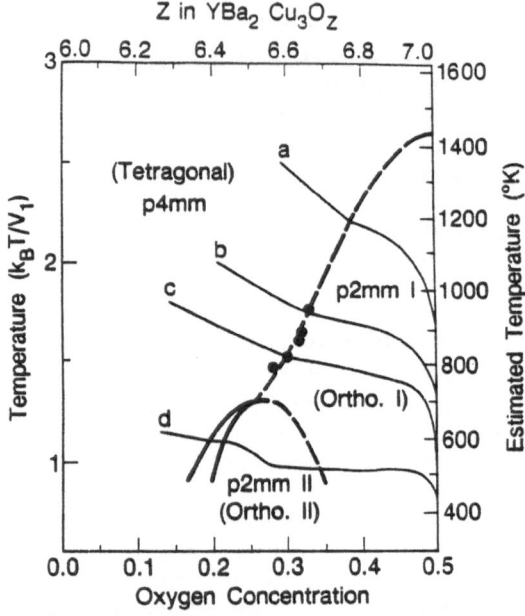

Fig. 4.   Phase diagram calculated for x=−0.75, y=+0.50. Dashed lines are second order transitions, full lines are first order phase boundaries. Filled circles are order-disorder transition points determined experimentally (Ref. 25).

atm. was placed on the calculated order-disorder transition line, thereby fixing the temperature scale, i.e. fixing the value of the first EPI V_1. All other points at 5×10^{-3}, 10^{-1}, 2×10^{-1} and 1 atm. then fell very nicely on the calculated curve. By this method, an estimated temperature scale could be constructed (right hand scale). An oxygen content scale, measured by the oxygen stoichiometry index z, is also shown at the top portion of Fig. 4. Oxygen partial pressure curves are also shown in Fig. 4; these were calculated by making use of two points from the Oak Ridge data, as explained elsewhere[24].

Very recently, Kikuchi and Choi[26] have extended to low temperatures our previous phase diagram calculation[27] for the case $V_2 = -0.5V_1$, $V_3 = +0.5V_1$ (point #3 in Fig. 2), thereby showing that our conjectured low temperature phase boundaries were incorrect. The Kikuchi-Choi (KC) phase diagram is reproduced in Fig. 5 (redrawn with our earlier temperature scales[27]) with four of the Oak Ridge data points shown. The superconductivity transition temperature as a function of oxygen content is also indicated (lower dashed curve). As in the earlier calculation[27], the line of second order Tetragonal (T) to Orthorhombic (OI) transition terminates in a bicritical point below which the Ortho. II (OII) phase is stable. The OI↔ OII transition is found to be second-order down to absolute zero of temperature while the T ↔ OII transition is first order. The upper dashed line is

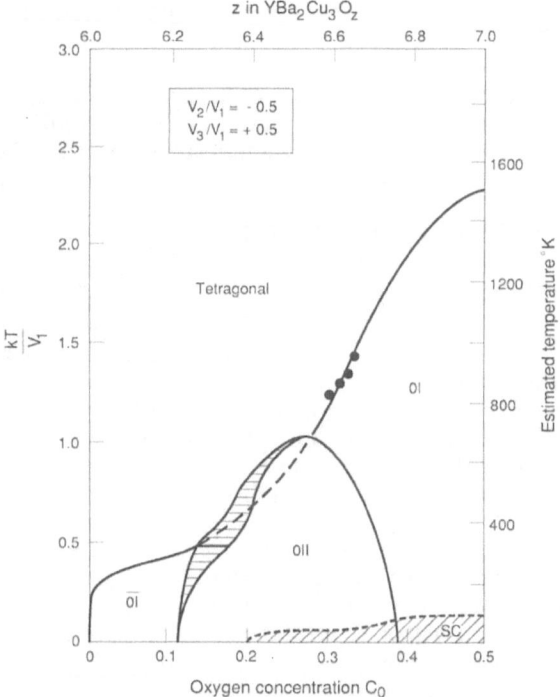

Fig. 5.   CVM phase diagram according to Kikuchi and Choi (Ref. 26) calculated
for x = –0.50, y = +0.50. Filled circles are order-disorder transition
points determined experimentally (Ref. 25)

the metastable extension, though the OII phase region, of the T ↔ OI line of
second order transitions. This line emerges from the OII region at a critical end
point to produce an "anti-Ortho. I" ($\overline{OI}$) phase region at low temperatures and
oxygen concentrations. The narrow two-phase region to the left of OII (horizontal
shading) appears to terminate at a tricritical point at low temperature.

The OI phase has the well-known "chain" structure (lower left sector of Fig.
3a) with occasional missing oxygen rows. The $\overline{OI}$ phase is, in a sense, the
converse: now the concentration of missing rows is much greater that that of filled
rows. Between OI and $\overline{OI}$, OII is characterized by double-cell long range order
(LRO) (upper left panel of Fig. 3b: every other chain missing). Phase $\overline{OI}$ differs
from the Tetragonal (T) in that the former has parallel chain LRO on one sublattice
only, T has fluctuating short chains on both oxygen sublattices, therefore along two
orthogonal directions; T has tetragonal, $\overline{OI}$ has orthorhombic symmetry. At low
enough temperature, an infinitesimal amount of oxygen will form long chains on a
single sublattice, thereby breaking the tetragonal symmetry. Recent transfer matrix
calculations[16] on the very same system indicate that the T ↔ OII transition is a
second-order one with critical exponents depending on concentration, as in the

Ashkin-Teller model. The other transitions are confirmed to be Ising-like and second-order.

The 60K superconductivity transition temperature plateau shown in Fig. 5 fits fairly well inside the OII phase boundaries at around 400K, where the oxygen configurations must be "frozen in". This suggests that OII is indeed the observed 60K superconductor.

ONE-DIMENSIONAL STATES OF ORDER

Monte Carlo simulations were also performed[28,29] with the same EPI ratios, i.e. those corresponding to point #3 on the parameter map of Fig. 2. At high temperatures, agreement with previous CVM calculations[12] was quite satisfactory. At low temperature simulations, around $\tau \equiv k_BT/V_1 = 0.2$ (k_BT has its usual "Boltzmann constant, absolute temperature" meaning), spurious specific heat maxima occurred in and around the OII phase field. Examination of computer printouts revealed that these specific heat anomalies corresponded to states of somewhat irregularly spaced parallel O-Cu-O chains.

A typical partially ordered structure is shown in Fig. 6a where black dots denote Cu atoms, shaded circles are oxygen atoms and open circles are vacant sites. This structure was obtained by performing a Monte Carlo simulation on a 64 x 64 square network of oxygen sites at fixed chemical potential field (normalized by V_1) of $\mu = -4.7$. Oxygen coverage in the plane is $c_o = 0.2265$ or stoichiometry index z = 6.453. The system was "quenched" from a high temperature of $\tau = 5.0$ and the iteration was pursued to 1200 Monte Carlo steps per site. O-Cu-O chains are seen to be fully formed and are stacked in such a way as to form regions of Ortho. II separated by slabs of "Ortho. III" (full-empty-empty-full cell tripling structure). The corresponding intensity pattern, i.e. the amplitude-squared of the Fourier transform, or oxygen structure factor of Fig. 6a, is shown in Fig. 6b. Diffraction maxima are located at <00>, <20> and <11> which are the "Bragg peaks" in planar reciprocal space notation, and at <10> which are the Ortho. I "reflections". The $<\frac{1}{2}0>$ reflections are split into satellite peaks at $<\frac{1}{2}\pm q,0>$, with q not necessarily a simple fraction. At an earlier stage of the simulation (500 Monte Carlo steps/site), the chains were often faulted and both α and β oxygen sublattice occupancy domains were observed (see Fig. 7). Chain formation dynamics is discussed in more detail by Burmester and Wille.[29]

Another mode of "sample preparation" also was used in the simulations: first Ortho. I was produced at $\mu = -2.0$, then oxygen was extracted by imposing a chemical potential field of $\mu = -3.2.$, this procedure being equivalent to reducing the partial pressure of Oxygen. After 8000 Monte Carlo steps/site, a structure consisting of mixed Ortho. II and Ortho. I slabs was obtained as seen in Fig. 8. The average planar oxygen concentration was $c_o = 0.297$, corresponding to stoichiometry z = 6.594. Once again the chains were fully formed but the OI and OII mixing was rather irregular, giving rise to a diffraction pattern consisting of a diffuse streak centered on $<\frac{1}{2}0>$ along the **a** direction. It is easy to rationalize how

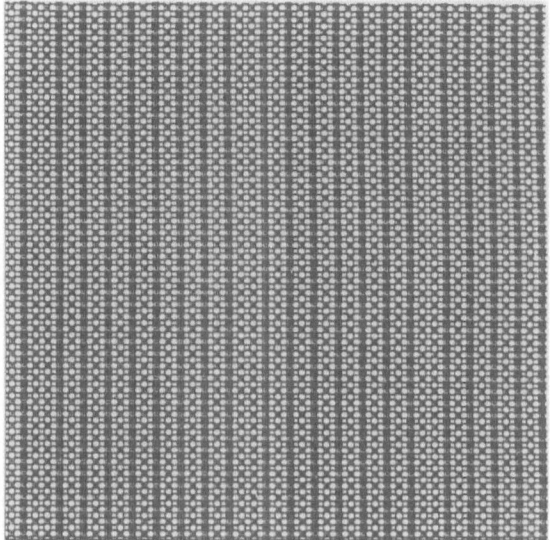

Fig. 6a. Chain configuration for Monte Carlo system quenched from $\tau = 0.2$ at constant $\mu = -4.7$, after 1200 Monte Carlo steps/site. Closed circles denote oxygen ions, open circles are vacant sites, and small dots (perhaps not readily visible) indicate Cu ions. Concentration is $c_0 = 0.2265$ or oxygen stoichiometry index $z = 6.453$. Structure is one of mixed OII and $\overline{OI}$ domains.

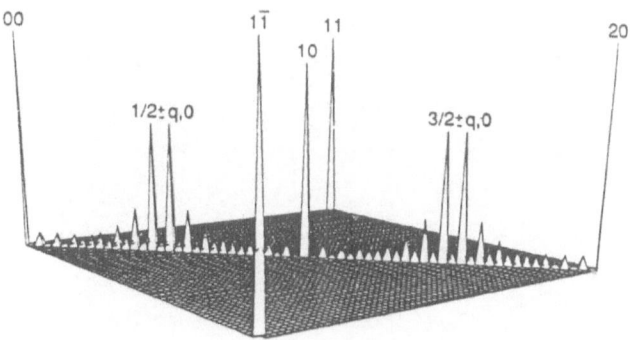

Fig. 6b. Fourier transform (amplitude squared) of configuration of Fig. 6a showing characteristic split (1/2, 0) peaks and (1,0) streaks of intensity.

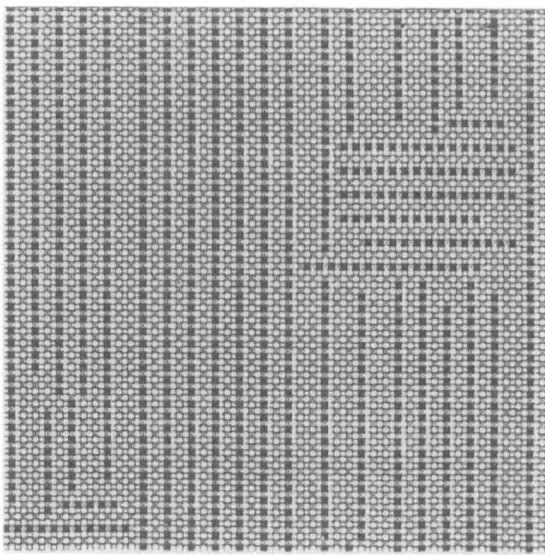

Fig. 7. Chain configuration for Monte Carlo system of Fig. 6a but at an early stage of iteration (500 Monte Carlo steps/site).

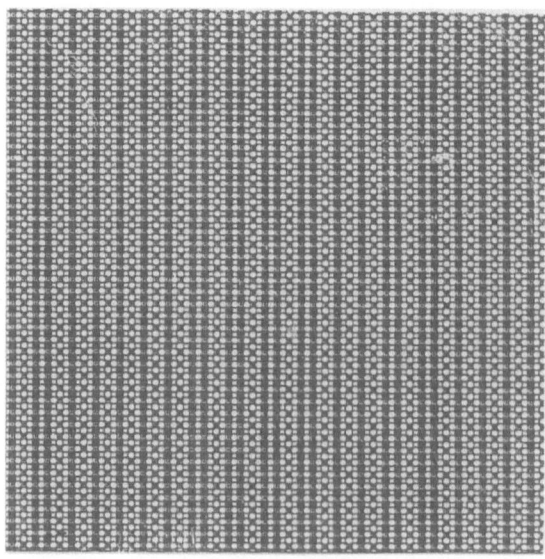

Fig. 8. Chain configuration for a Monte Carlo system initially in Ortho. I state, equilibrated at $\mu = -3.2$. Configuration is $c_0 = 0.297$ or $z = 6.594$. Structure is one of mixed OI and OII domains.

such a structure could come about. Because of periodic boundary conditions, complete O-Cu-O chains are infinite and thus very robust. Eventually a chain is broken by action of the chemical potential field and it subsequently dissolves, thereby creating a slab of Ortho. II. The process is then repeated but in such a way that, at least for $c_0 > 0.25$, no two empty chains form adjacent to one another. The reason for that is as follows: fully formed chains on a single sublattice interact only through the effective V_3 interaction which is repulsive ($V_3 > 0$), thus favoring unlike nearest neighbor parallel chains. Hence a mixed state of OI and OII domains results. For $c_0 < 0.25$ ($z < 6.5$), the situation is slightly different: the mixing is between OII and $\overline{OI}$ domains.

This quasi one-dimensional admixture of domains, OII + OI for $c_0 > 0.25$, OII + $\overline{OI}$ for $c_0 < 0.25$, gives rise, locally, to structures resembling the Magneli phases described by Khachaturyan and Morris (KM). Superficially, it may appear that the present findings confirm the KM model of "transient homologous structures". In point of fact, quite the opposite is true, as will now be shown. First note that the "method of concentration waves" as used by KM to construct the reported phase diagrams[30,31] here reduces exactly to a two dimensional Bragg-Williams model with first and second neighbor effective pair interactions. A simple back-Fourier transform of the KM mean field internal energy, for their chosen star values of $V(0)$ and $V(k_1)$, immediately yields the required values (in our notation) of V_1 and V_2 (=V_3), with ratio $V_2/V_1 = -0.4286.....$The representative point of this ratio is indicated by a cross (x) on the diagonal in Fig. 2, near point #1. This ratio, inserted into the Bragg-Williams model, as expected, reproduced exactly the KM phase diagram.[32]

Next consider the "Magneli" structures proposed by KM: $[(O\square)^j O]$ where O and $\square$ represent filled and empty chains on the α sublattice respectively, where the square brackets denote periodic repetition and where the parentheses indicate that the Ortho. II element is to be repeated j times. Such structures are actually a subset of the well-known ground states of the one-dimensional Ising model and are stabilized by dominant *positive* (i.e. repulsive) first neighbor pair interaction J_1 along the Ising line[33,34] which is the **a** direction in orthorhombic 1-2-3. But, as mentioned above, for fully formed chains the only relevant interaction is the interchain repulsive interaction V_3 which is precisely J_1. Hence the presence of "Magneli" structures, stable or metastable, necessarily requires $J_1 \equiv V_3 > 0$. The KM phase diagram, however, is constructed with $V_2 = V_3 < 0$ so that this model is actually internally inconsistent, as was already noted earlier[4]. A check was performed by a Monte Carlo simulation employing the "symmetric" KM interaction parameters $V_2 = V_3 = -0.4286V_1$ and, as expected, no Magneli-like ordered states, stable or otherwise, were found. From the foregoing it is thus clear that *the presence of states of partial one-dimensional order necessarily implies that Ortho. II must be a stable phase in the system.*

At low temperature, the behavior of the proposed model with $V_2 > 0$ and $V_3 < 0$ is highly unusual: at $\tau \approx 0.2$, the interchain O $-$ O correlation length is expected to become greater than the dimension of the Monte Carlo periodic cell so that spurious results may occur as discussed for example by Selke[35] in the context of the two dimensional ANNNI model. The system essentially can be mapped onto a one-dimensional Ising model with nearest neighbor J_1 interaction for which only three ground states exist: the "ferromagnetic" (OI and $\overline{OI}$) and the "antiferromagnetic" (OII). Moreover, for $J_1 \equiv V_3 > 0$, any mixing of "ferro." and "antiferro." domains at fixed average O-concentration yields the same energy so that complete degeneracy prevails.

The mixing of domains appears to be not completely random, however. Indeed, at intermediate temperatures when the full two-dimensional character of the system is recognized by the Monte Carlo simulation, OI, $\overline{OI}$ or OII long range order (LRO) is observed at equilibrium, as expected from the phase diagram. At low-temperature, the two-dimensional character of the model is not totally absent, especially during the early stages of chain formation (see Fig. 7) so that effective long-range (entropy driven) interchain repulsion may occur. Resulting structures and their Fourier transforms are clearly reminiscent of ordered structures and diffraction patterns observed from 1-2-3 samples annealed at low temperatures for long periods of time under reducing conditions with final oxygen stoichiometry between about $z \equiv 6.25$ and 6.75 [36]: split diffraction peaks and diffuse streaks are seen about the [1/2,0] positions, but rather sharp peaks at positions [1/3,0] and [2/3,0] indicate, however, that actual longer-range repulsive interactions, beyond V_3, may be present along the **a** direction.

DISCUSSION

Results from phase diagram calculations and Monte Carlo simulation appear to be in close agreement with experimental findings. Indeed, at this same conference, Professor Amelinckx[37] showed some remarkable high resolution electron micrographs and diffraction patterns which confirm the general picture which emerges from theoretical findings. The close agreement was undoubtedly facilitated by the Antwrp group's[37] use of a sample preparation technique featuring very slow cooling of $YBa_2Cu_3O_z$ samples *at constant oxygen content z* (through use of a feedback control on a thermogravimetric balance). In this way, low temperature equilibrium at fixed z could be achieved which previous quenching methods could not. With this technique, one-dimensional OI, OII and $\overline{OI}$ domains were seen in the micrographs[37], just as observed in the simulations.

Moreover, Ortho. II was clearly shown to be a stable phase. In earlier dark field electron microscopy work on quenched samples[11], only small volume fractions of Ortho. II were reported, but it was argued elsewhere[4,22] that this was probably the result of poor correlation of mirror planes along the **c** direction: OII domains phase shifted from plane-to-plane will project as OI thereby causing appreciable underestimation of the areal fraction of OII on each plane. With constant-concentration slow cooling, however, the plane-to-plane correlation could be made almost perfect with only occasional antiphase shifts[13]. This result may appear somewhat surprising in view of the fact that O–O effective interactions between mirror planes along **c** must be heavily screened by intervening ions. Close to the critical point for two dimensional ordering however, correlation lengths in each plane become infinite so that effective ordering interactions between planes may become arbitrarily large with respect to the one-dimensional faulting tendency due to entropy[17]. Thus, by slow constant-z cooling, nearly perfect three-dimensional OII ordering may be attained. For Ortho. I there is no problem: elastic strains due to orthorhombicity make the exchange of **a** and **b** axes orientations far too costly.

The asymmetry of the proposed Ising model ($V_2 < 0$, $V_3 > 0$) guarantees the formation of very long chains along the **b** axis at low temperature, thereby leading to quasi one-dimensional states of order. High one-dimensional degeneracy also results at very low temperature as demonstrated by Monte Carlo simulation, and in agreement with the low temperature CVM calculation of Kikuchi and Choi[26]: it is seen in Fig. 5 that all low temperature transitions are second order so that a complete series of domain mixing can obtain, in no violation of the third law of thermodynamics, recent claims to the contrary notwithstanding[30,31].

The Kikuchi-Choi phase diagram appears to be the most complete to date although there are indications that more complex ordered phases may become stable at low temperatures[8,36]. Indeed, recent Monte Carlo simulations performed with $V_4/V_1 = 0.25$ (V_4 being the next nearest interchain interaction, beyond V_3) lead to stability of OO□ and O□ □ "Magneli" phases[39]. It is important to note, however, that the existence of these and other states of order is incompatible with "symmetric" ($V_2 = V_3$) miscibility gap models. With large enough V_4 (and beyond), the phase diagram of Fig. 5 of course would have to be modified.

ACKNOWLEDGEMENTS

Ground state and phase diagram calculations were performed in collaboration with A. Berera and L. T. Wille. Monte Carlo studies were performed by C.P. Burmester, M.E. Mann and L.T. Wille with programming assistance from C. Carter. Special thanks are due to Dr. R. Kikuchi who allowed the author to reproduce the phase diagram of Fig. 5. Helpful discussions were held with S. Amelinckx, G. Ceder, D.H. Lee, S. C. Moss, J. D. Jorgensen and H. Zandbergen. The research was supported by a grant from the Director, Office of Energy Research, Materials Sciences Division, U.S. Department of Energy, under contract DE-AC03-76SF00098.

REFERENCES

1. M. K. Wu, J. R. Ashburn, C. J. Torny, P. H. Hor, R. L. Meng, L. Gao, Z. J. Huang, Y. Q. Wang and C. W. Chu; Phys. Rev. Lett. 58, 908 (1987).
2. R. Beyers and T. Shaw, in "Solid State Physics," Vo. 42, H. Ehrenrich and D. Turnbull, Eds., pp. 135-212, Academic Press, NY (1989).
3. D. de Fontaine, L.T. Wille and S.C. Moss, Phys. Rev B 36, 5709 (1987).
4. D. de Fontaine, in " Proceeding of Fifth Meeting of Japanese Committee for Alloy Phase Diagrams," June 1988.
5. F. Ducastelle and F. Gautier, J. Phys. F 6, 2039 (1976).
6. L.T. Wille and D. de Fontaine, Phys. Rev. B 37, 2227, (1988).
7. G. Van Tendeloo, H.W. Zandbergen and S. Amelinckx, Sol. St. Comm. 63, 603 (1987).
8. M.A. Alario-Franco, J.J. Capponi, C. Chaillout, J. Chenevas and M. Marezio, Mat. Res. Soc. Symp. Proc. 99 (1988).

9. R.J. Cava, B. Batlogg, C.H. Chen, E.A. Rietman, S.M. Zahurak and D. Werder, Phys. Rev. B 36, 5719 (1987).
10. R.M. Fleming, L.F. Schneemeyer, P.K. Gallagher, B. Batlogg, L.W. Rupp and J.V. Waszczak (preprint).
11. C.H. Chen, D.J. Werder, L.F. Schneemeyer, P.K. Gallagher and J.V. Waszczak (preprint).
12. Y. Nakazawa, M. Ishikawa, T. Katabatake, H. Takeya, T. Shibuya, K. Terakura and F. Takei, Japanese J. Appl. Phys. 26, L682, (1987); Y. Nakazawa, M. Ishikawa, T. Katabatke, K. Koga and K. Terakura, Japanese J. Appl. Phys. 26, L798 (1987).
13. S. Amelinckx, Proceedings of this conference.
14. F. Claro and V. Kumar; Surf. Sci. 119, L371 (1982).
15. K. Binder and D.P. Landau, Phys. Rev. B 21, 1941 (1980).
16. N.C. Bartelt, T.L. Einstein and L.T. Wille (preprint).
17. D.H. Lee, private communication.
18. J. Ihm, D.H. Lee, J.D. Joannopoules and J.J. Xiong, Phys. Rev. Letters, 51, 1872 (1983)
19. Per Bak, P. Kleban, W.N. Unertl, J. Ochab, G. Akinci, N.C. Bartelt and T.L. Einstein, Phys. Rev. Letters 54, 1539 (1985).
20. R. Kikuchi, Phys. Rev. 81, 988 (1951).
21. A. Berera, L.T. Wille and D. de Fontaine, J. Stat. Phys.: Short Communication 50, 1245 (1988).
22. D. de Fontaine, in "Proc. of NATO Institute on Phase Stability," Crete, July 1987 (in press).
23. J. C. Wheeler, private communication.
24. A. Berera and D. de Fontaine, Phys. Rev. B 39, 6727 (1989).
25. E.D. Specht, C.J. Sparks, A.G. Dhere, J. Brynestad, O.B. Cavin, D.M. Kroeger and H.A. Oye, Phys. Rev. B 37, 7426 (1988).
26. R. Kikuchi and J.S. Choi (preprint).
27. L.T. Wille, A. Berera and D. de Fontaine, Phys. Rev. Lett. 60, 1065 (1988).
28. D. de Fontaine, M.E. Mann and G. Ceder, to be submitted.
29. C.P. Burmester and L.T. Wille (preprint).
30. A.G. Khachaturyan and J.W. Morris, Jr., Phys. Rev. Lett. 61, 215 (1988).
31. A.G. Khachaturyan and J.W. Morris, Jr., Phys. Rev. Lett. 59, 2776 (1987).
32. G. Ceder, unpublished work at U.C. Berkeley (1989).
33. S. Katsura and A. Narita, Prog. Th. Phys. 50, 1750 (1973); M. Kaburagi and J. Kanamori, Prog. Th. Phys. 54, 30 (1975).
34. A. Finel, These de Doctorat d'Etat, Universite Pierrre et Marie Curie, July 1987 (unpublished) and references cited therein.
35. W. Selke, Z. Physik B - Condensed Matter 43, 335 (1981).
36. R. Beyers (private communication).
37. M.E. Mann, unpublished work at U.C. Berkeley (1989).

EFFECTS OF CORRELATION AND DISORDER ON THE ELECTRONIC

PROPERTIES OF $RBa_2 Cu_3 O_{6+x}$

A. Latgé, E.V. Anda and J.L. Morán-López*

Universidade Federal Fluminense
Outeiro de São João Batista SN
Niteroi, Río de Janeiro, Brasil
*Universidad Autónoma de San Luis Potosí
78000 San Luis Potosí, México

INTRODUCTION

The study of the crystallographic structure of high T_C super-conducting materials has been crucial to understand its supercon-ductive properties. Neutron power diffraction studies [1] have revealed that the $YBa_2Cu_3O_{6+x}$ system is an oxygen deficient perov-skite with two CuO_2 planes and a CuO_x (0<x<1) plane, which for x=1 and at low temperatures consists of linear chains of Cu and O atoms.

Several band structure calculations of $RBa_2Cu_3O_{6+x}$ using different local density functional approximation schemes [2,3] have revealed that the main features of the band structure at the Fermi level can be obtained by considering only two electronic orbitals to describe the chains and the planes; the d and p orbitals associated to Cu and O respectively.

In a previous work [4] we used a simple tight-binding ap-proach, suggested by the local density functional calculations, to study the effect that the oxygen disorder in the basal plane pro-duces in the electronic density of states of the compound. However, a severe shortcoming of that study, as well as that of the local density functional schemes, was that they did not included in an adeauate form the electronic interactions of a local highly corre-lated system, as it is believed to be the case of high T_C ceramics because of the large intra-atomic electronic repulsion at Cu and O sites.

A proper understanding of superconductivity in $RB_2Cu_3O_{6+x}$ requires a detailed analysis of the simultaneous effects that electronic correlations and disorder have on the region near the Fermi level and in particular its influence on the equilibrium interchange of electronic charge (as a function of doping) between the localized linear chain states and the extended states of the CuO_2 planes.

The purpose of this paper is to treat the high correlated disordered electronic gas in a simple way but accurate enough to permit an adequate description of the Mott -Anderson metal-non-metal transition that this system suffers as a function of the oxygen content. Simultaneously, we study the charge redistribution between the CuO_2 planes and the chains and the general electronic properties of the system.

Although many efforts have been devoted to understand $RBa_2Cu_3O_{6+x}$ there are many open questions concerning the essential mechanism for superconductivity, as well as in relation to the role of the different oxygen and copper atoms, their electronic charge and their crystallographic structure in the basal and CuO_2 planes. There is as well a lack of a microscopic model for high T_c oxides for $T>T_c$ which we believe could help to understand the superconductive phase.

To describe the $RBa_2Cu_3O_{6+x}$ systems we propose a Hubbard type Hamiltonian, which includes the intrasite Cu and O and the nearest-neighbors intersite O-O and O-Cu electronic repulsions.

The Hamiltonian that we consider is the following:

$$H = H_{chain} + H_{plane}$$

$$H = \sum_{\beta i \sigma} E^{\beta}_{0\sigma} c^{}_{i\sigma} c^{}_{i\bar{\sigma}} + \sum_{\beta \alpha \sigma} E^{\beta}_{Cu} d^{+}_{\alpha\sigma} d^{}_{\alpha\bar{\sigma}} + U_p \sum_{i\sigma} n_{i\sigma} n_{i\bar{\sigma}} + \tag{1}$$

$$+ U_d \sum_{\alpha\sigma} N_{\alpha\sigma} N_{\alpha\bar{\sigma}} + U_{pd} \sum_{\substack{i\alpha\sigma \\ \sigma'}} n_{i\alpha} N_{\alpha\sigma'} + U_{pp} \sum_{\substack{ij\sigma \\ \sigma'}} n_{i\sigma} n_{j\sigma'} +$$

$$+ \sum_{i\alpha\sigma} V^{i\alpha}_{pd} (c^{+}_{i\sigma} d^{}_{\alpha\sigma} + c.c.)$$

with: $n_{i\sigma} = c^{+}_{i\sigma} c_{i\sigma}$, $N_{\alpha\sigma} = d_{\alpha\sigma} d^{+}_{\alpha\sigma}$ and where $i(\alpha)$ denotes an oxygen (copper) atom. The $c^{+}_{i\sigma}(d^{+}_{\alpha\sigma})$ operators create an electron in site $i(\alpha)$ with spin σ at the p and d orbitals and β corresponds to the two: chains or planes. U_p, U_d, U_{pd} and U_{pp} are the intra and intersite correlations. The matrix element $V^{i\alpha}_{pd}$ corresponds to the hopping between copper at site α and the nearest-neighbour oxygen atoms at sites i.

THE ELECTRONIC DENSITY OF STATES

The solution of the many body Hamiltonian (1) describing a system of randomly distributed atoms (x<1) requires the simultaneous treatment of disorder and electronic correlation. The many body problem was considered using the alloy analogy approximation for the intra-atomic repulsion at Cu and O sites [5], and a Hartree Fock approach for the smaller intersite O-O and O-Cu repulsions ($U_d \approx 9.0$ eV; $U_p \approx 6.0$ eV; $U_{nd} \approx 1.0-6.0$ eV and $U_{pp} \approx 0.7-1.6$ eV, following ab initio calculations, [6,7]). To obtain the quantum mechanical and the configurational averages of the relevant physical variables we adopted the Green function method. The diagonal Green functions are calculated applying a real space renormalization technique which decimates interatively the lattice points corresponding to Cu and O sites. At each stage of decimation, the eliminated degrees of freedom are configurationally averaged, preserving the statistical correlations between diagonal and non diagonal renormalized parameters [8]. The existence of the O(4) atoms linked to the coppers is taken into account by renormalizing the Cu(1) undressed diagonal Green function. The system of equations of motion for the Green function $G_{\alpha\beta}(\omega)$, corresponding to a semi-infinite linear chain is:

$$G_{00}(\omega) = g_0^\gamma + g_0^\gamma \, V_{0,1} \, G_{10} \qquad\qquad (2)$$

$$G_{10}(\omega) = g_1^\alpha \, V_{1,0}^0 \, G_{00}(\omega) + g_1^\alpha \, V_{1,2}^0 \, G_{20}(\omega) \qquad\qquad (3)$$

$$G_{20}(\omega) = g_2^\beta \, V_{2,1}^{Cu} \, G_{10}(\omega) + g_2^\beta \, V_{2,3}^{Cu} \, G_{3,0}(\omega) \qquad\qquad (4)$$

$$\ldots \qquad \ldots\ldots\ldots\ldots\ldots\ldots\ldots\ldots\ldots\ldots\ldots\ldots$$
$$\ldots \qquad \ldots\ldots\ldots\ldots\ldots\ldots\ldots\ldots\ldots\ldots\ldots\ldots$$

where the undressed Green function is given by: $g_i = \dfrac{1}{\omega - E_i^\lambda}$; where $\omega = \omega + i\eta$ and $\eta = 0^+$

The subindex $i(\alpha)$ is associated to the O(Cu) sites, while λ denotes single or double electronic occupied atoms. For the sake of simplicity, we describe the CuO_2 planes by a copper-oxygen decorated Bethe lattice. For the CuO_2 plane, within the Bethe lattice approximation, the decimation is applied to equations (2-4) where the renormalized diagonal Cu(1) Green functions are replaced by the corresponding undressed Cu(2) Green functions.

The linear chains in the basal plane are responsible for the spatial disorder introduced into the eigenstates near the Fermi level. We will assume that reducing the oxygen content the vacancies will appear at random along the well established linear chains. After the decimation procedure of the dilute chain has been done,

we consider three different clusters for which Cu(1) is fourfold,
threefold and twofold coordinated with an statistical weight of
x^2, $2(1-x)x$ and $(1-x)^2$ respectivelly, in order to take into account
the local environment disorder.

The position of the Fermi level and the occupation numbers
are obtained self-consistently by solving the equations:

$$2 \int_{-\infty}^{Ef}(2\rho(\omega)_{Cu(2)} + 4\rho(\omega)_{O(2)} + 2\rho(\omega)_{O(4)} + \rho(\omega)_{Cu(1)} + x\rho(\omega)_{O(1)})\ d\omega = 9+Q \quad (5)$$

and $$2 \int_{-\infty}^{Ef} \rho(\omega)_{local}\ d\omega = n_i \qquad\qquad (6)$$

where $\rho(\omega)$ is the local densities of states of each atom and Q is
the cationic charge of the noncopper metal transfered to the
planes and chains which is equal to 7 per unit cell for the case
of $YBa_2Cu_3O_{6+x}$. The number 2 in front of the integrals results
from spin degeneracy. We were able to investigate the electronic
behaviour of different compounds and in particular the metal-non
metal transition as a function of the oxygen content and the
transfered charge Q.

RESULTS AND DISCUSSION

We show in figure 1(a and b) the total density of states of
$YBa_2Cu_3O_{6+x}$ ($x=1$, and 0.7, respectively) calculated by solving the
Hamiltonian (1). The intra atomic electronic repulsion at the
copper and oxygen atoms has split the states into a sort of lower
and upper Hubbard bands [9]. However, the correlation gap is more
complicated than what the Mott Hubbard model could predict. It is
a charge transfer gap existing essentialy between the filled upper
Hubbard O 2p-band and the empty upper Hubbard 3d-band. This model
predicts the energy level to be near the correlation gap. This
could explain why in photoemission spectra [10] it is observed a
LDOS at the Fermi level, much smaller than the results obtained by
the density functional calculations.

We have calculated the local density of states (LDOS) of Cu
and O as a function of the oxygen content as can be seen in Figure
2(a-c) and 3(a-d). The strong intra-atomic electronic repulsion of
Cu, U_d, splits the states far apart from the Fermi level, and the
O(2) and O(1) bands, less splitted by the electronic repulsion,
reveal the predominant oxygen 2p character of the states near the
Fermi level. This is in agreement with several spectroscopy experiments [11,12] showing the great importance of considering local
correlations to get an appropriate description of the material.
The LDOS of Cu(1) and O(1) exhibit a strong dependence upon oxygen
content x, due to the direct disorder that the oxygen vacancies
introduce into the Cu-O chain, while the CuO_2 plane LDOS is almost
independent of x.

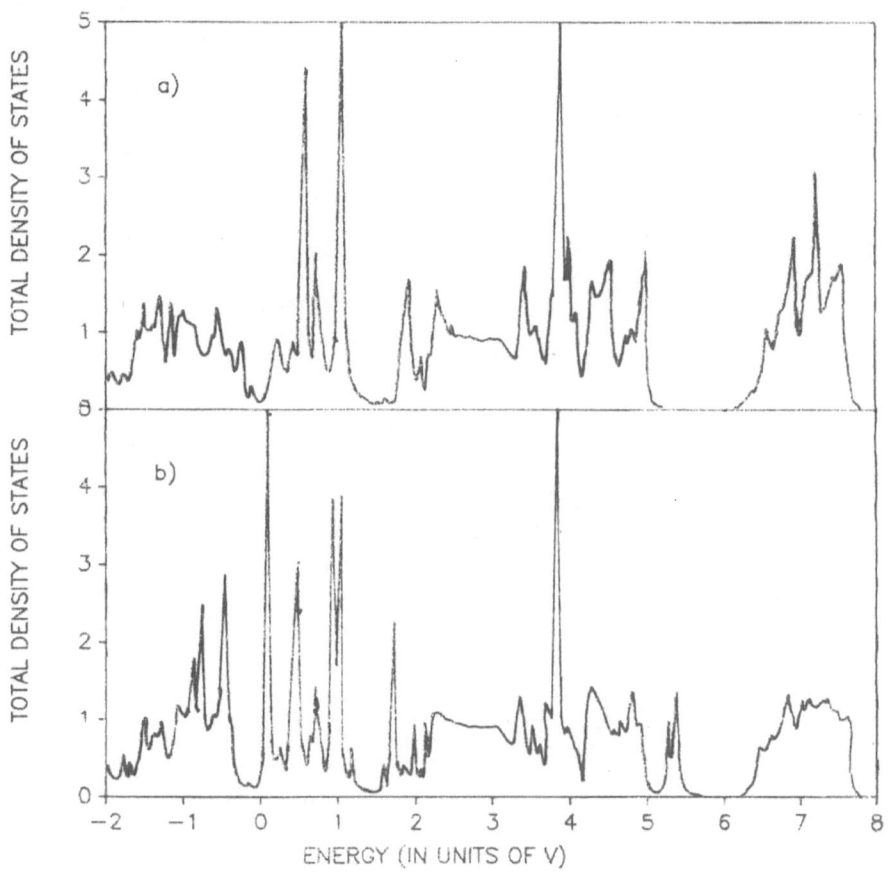

Fig. 1.   Total LDOS for a) YBa$_2$Cu$_3$O$_7$ and b) YBa$_2$Cu$_3$O$_{6.7}$

The LDOS of the system shows a well defined gap for x=1, which is filled by an increasing, however small, amount of states as the number of oxygen vacancies increases. They correspond to O(1) and Cu(1) Hubbard eigenvalues that are shifted to higher and lower energy values, respectively, due to changes in the nearest environment of these atoms produced by dilution. The oxygen vacancies creates twofold and threefold coordinated coppers which are able to bind electrons more strongly than fourfold coordinated coppers, since its electrons do not suffer the repulsion of the electronic charge at the oxygen sites which are missing. In figure 4 we present the dependence of the Fermi energy as a function of the oxygen content. For values of x for which the Fermi level lies at the upper oxygen band, the variation is almost linear. As soon as E$_f$ enters into the pseudogap it shows a rapid increase. This results compares well with photoemission experiments, at least for the range 0.2<x<1 for which we get an energy difference of 0.72 eV,

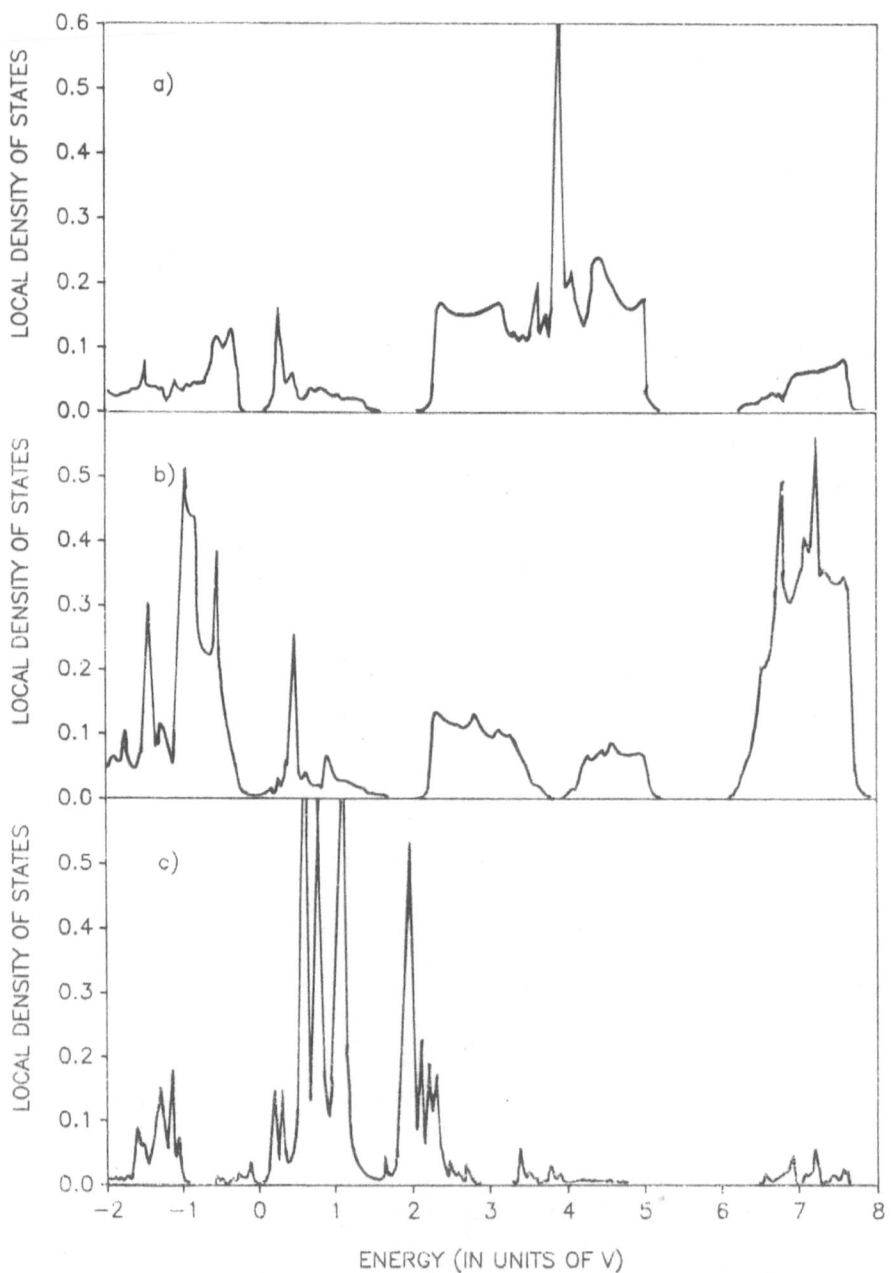

Fig. 2. LDOS for the case of x=1: a) O(2); b) Cu(2) and c) O(4)

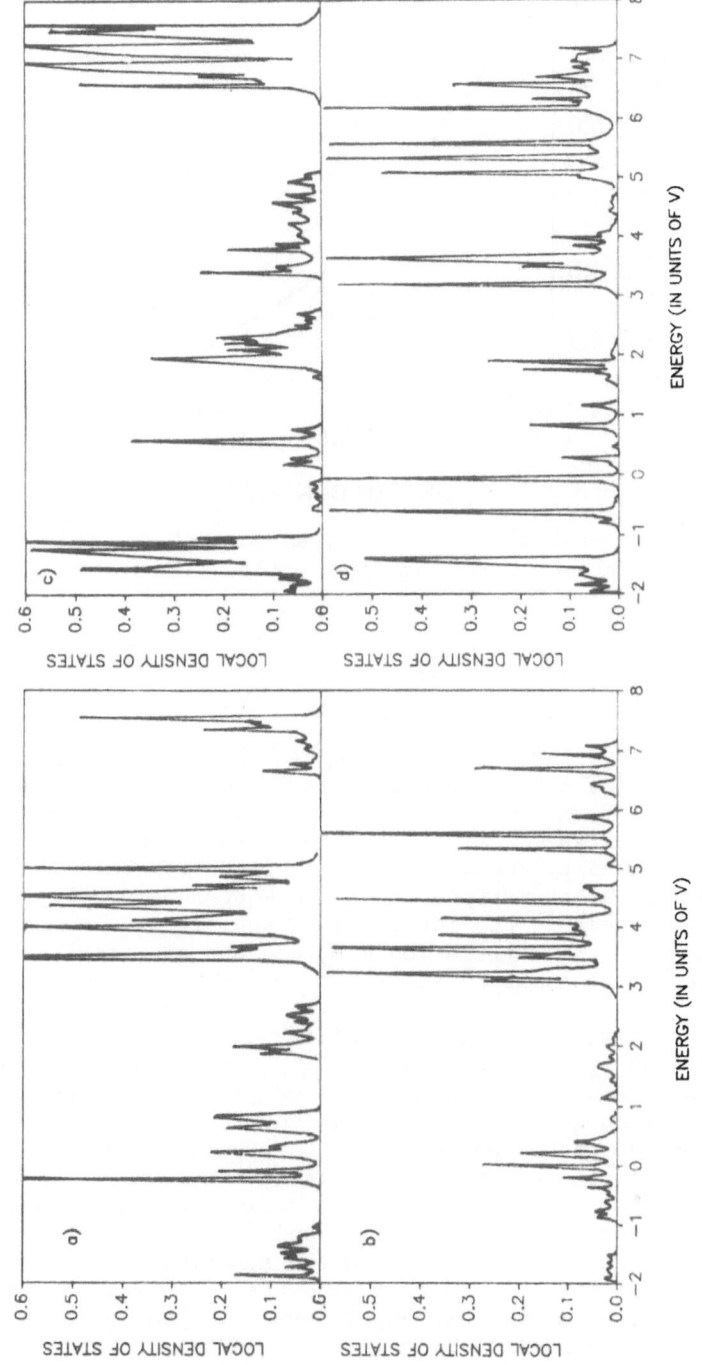

Fig. 3. LDOS for two different oxygen concentrations. Oxygen LDOS are shown on 3a) x=1 and 3b) x=0.5. Copper LDOS are shown on 3c) x=1 and 3d) x=0.5

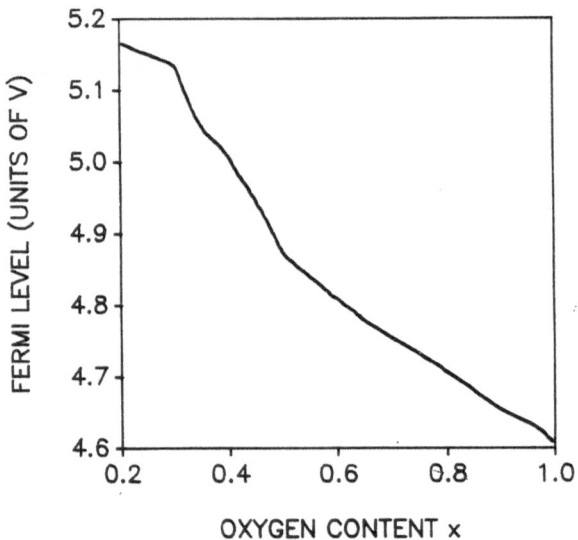

Fig. 4.   Fermi level position vs oxygen content for the case of
Q=7.0 electrons per unitary cell.

compared with the experimental result of 0.90 eV [13]. The system
suffers a metal-non metal transition as a function of the oxygen
content when the Fermi level goes through the delocalized states
of the CuO_2 plane (essentialy $O(2)$ states) and enters into the
pseudogap populated by localized Cu-O states. From this view point
the metal-non metal transition is of the Mott-Anderson type [14]
since it is produced by the interplay of the electronic correlation, which opens the correlation gap, and the disorder, which
populates it with localized states.

Within the range of solid solubility of the compounds
$Y(La_yBa_{1-y})_2Cu_3O_{6+x}$ and $(Y_yCa_{1-y})Ba_2Cu_3O_{6+x}$ the metal-non metal
transition has been experimentaly studied [15] as a function of
the oxygen content x and changing the concentration Y as a function
of the cation charge Q transfered to the planes and chains. We
were able to analyse this problem changing the value of Q in formula (5). Figure 5 shows, in the space defined by the parameters
x and Q, the theoretical (solid) line and the experimental (dashed)
line, separating the insulating from the superconducting phase.
The theoretical results agree well with the relatively disperse
experimental measurements that depend upon the procedure in which
the sample is prepared [16].

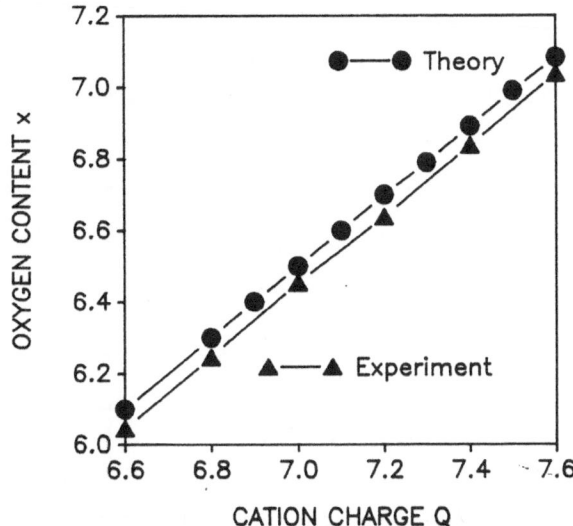

Fig. 5. Metal-non metal transition for a range of $YBa_2Cu_3O_{6+x}$
compounds. Solid line correspond to the theoretical
results while the dash line is from Ref. 15.

        The determination of the valence states of Cu (Cu^{3+}, Cu^{2+} and
Cu^{1+}) in these systems has provided up to now, confusing results
in relation not only to the existance Cu^{3+} state 17 , but also to
the role that these distinct charge states play in the mechanism
of superconductivity. We calculated the occurrence probabilities
of the three charge states for Cu in the CuO_2 planes and in the
Cu-O chains, as a function of the oxygen content x. The results
are shown in figures 6(a-b). The correlation $\langle n_i\, n_{i\bar{\sigma}}\rangle$, which is
equal to the probability of having two electrons in a particular
atom at site i can be obtained by integrating the imaginary part
of the diagonal Green function component associated to double
electronic occupation.

        A remarkable difference between the behavior of the Cu(2) and
Cu(1) charge probabilities can be observed. There is a clear reduction of Cu^{++} and a large increase of Cu^{+} as the oxygen content is
reduced. This result can be understood in view of the increase of
twofold coordinated Cu(1) generated by vacancies 18 . On the
other hand the Cu(2) valence does not depends in any significantly
way upon the oxygen content.

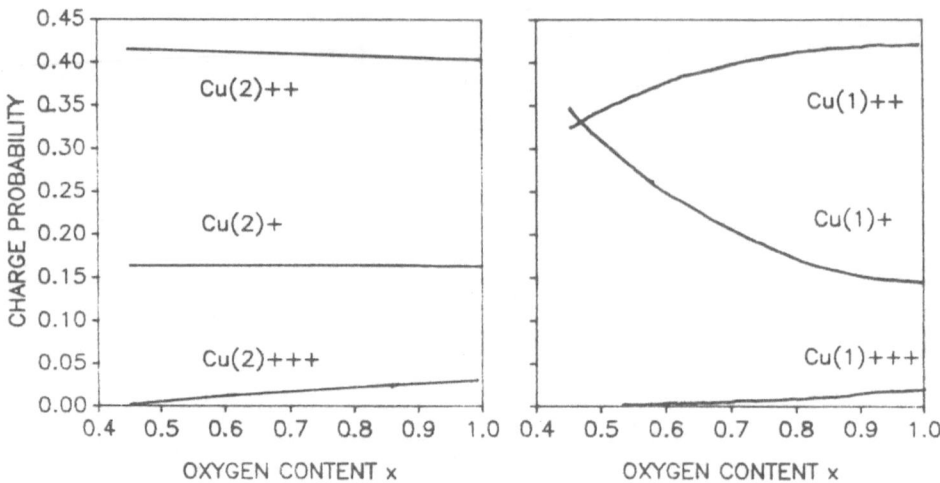

Fig. 6. Calculated charge probability of the three charge states
(Cu^+, Cu^{++} and Cu^{+++}) for the copper atoms of the a) CuO_2
plane and b) Cu-O chain as a function of oxygen content.

ACKNOWLEDGEMENTS

      This work was partially supported by DGICSA-SEP throgh Grant
C88-08-0170 and by PRDCT-OEA.

REFERENCES

1. J.D. Jorgensen, M.A. Beno, D.G. Hinks, L. Soderholm, K.J.
    Volin, R.L. Hitterman, J.D. Grace, I.K. Schuller, C.U. Segre,
    K. Zhang and M.S. Kleefish, Phys. Rev.B 36, 3608 (1987)
2. L.F. Mattheiss, Phys. Rev. Lett. 58, 1028 (1987)
3. S. Massida, J. Yu, A.J. Freeman and D.D. Koelling, Phys. Lett
    A 122, 198 (1987)
4. E.V. Anda and J.L.M. López, Proceedings of the Brasilian
    Simposium on Physics on High Superconductor, Río de Janeiro,
    World Scientific, Singapore (1988)
5. J. Hubbard, Proc. R. Soc. A 277, 237 (1964)
6. M.Schluter, M.S. Hybertsen and N.E. Christensen, in Proceed-
    ings of the International Conference on High T Superconduc-
    tors: Materials and Mechanisms of Superconductivity, Inter-
    laken, Switzerland, 1988, edited by J. Muller and J.L. Olsen
7. C.F. Chen, X.W. Wang, T.C. Seung and B.N. Harmon (unpublished)
8. E.V. Anda and A. Latgé, J. Phys. C: Solid State Phys. 21,
    4251 (1988)
9. J. C. Fuggle, P.J.W. Weijs, R. Schoorl, G.A. Sawatzky, J.
    Fink, N. Nucker, P.J. Durham and W.M. Temmerman, Phys. Rev.B
    37, 123 (1988)

10. N. Nucker, J. Fink, B. Renker, D. Ewert, C. Politis, P.J.W. Weijs and J.C. Fuggle, Z. Phys. B 67, 9 (1987)

11. A. Fujimori, E. Takayama-Muromachi, Y. Uchida and B. Okai, Phys. Rev. B 35, 8814 (1987)

12. P. Steiner, S. Hufner, V. Kinsunger, J. Sander, B. Siegwart, H. Schmitt, R. Schulz, S. Junk, G. Schwitzgebel, A. Gold, C. Politis, H. P. Muller, R. Hoppe, S. Kemmler-Sack and C. Kunz, Z. Phys. B 69, 449 (1988)

13. N.G. Stoffel and J.M. Tarascon, Phys. Rev. B 36, 3986 (1987)

14. Sir Nevill Mott, Metal-Insulator Transitions, Taylor and Francis Ltd 10-14 Macklin Street, London WC2B 5NF, 1987

15. Y. Tokura, J. B. Torrance, T. C. Huang and A. I. Nazzal, Phys. Rev. B 38, 7156 (1988)

16. H. Nozaki, S. Takekawa and Y. Ishizawa, Jpn. J. Appl. Phys. 27, L 31 (1988)

17. A. Balzarotti, M. De Crescenzi, N. Motta, F. Patella and A. Sgarlata, Phys. Rev. B 38, 6461 (1988)

18. W.W. Waren, Jr., R.E. Walstedt, G.F. Brennert, R.J. Cava, B. Batlogg and L.W. Rupp, Phys. Rev. B 39, 831 (1989).

TIGHT-BINDING INVESTIGATION OF THE ELECTRONIC PROPERTIES OF ORDERED AND DISORDERED DEFECTS IN THE YBaCuO SYSTEM

Ph. Lambin*

Department of Physics
Facultés Universitaires Notre-Dame de la Paix
B 5000 Namur, Belgium

ABSTRACT

Using a semi-empirical tight-binding Hamiltonian and the recursion technique, the number on holes in the CuO_2 square-planar networks brought about by the Cu-O chains in $YBa_2Cu_3O_{7-\delta}$ are computed for δ varying continuously between 0.0 and 0.5, either for random occupancies of the crystallographic O sites or ordered O-vacancy distributions consistent with experimental observations. In this connection, the electronic structure of $YBa_2Cu_4O_8$ and the electronic properties of a (110) twin boundary in $YBa_2Cu_3O_7$ are also analyzed. It is shown that O-vacancy arrangements leading to intact Cu-O chains are responsible for increasing the hole count in the CuO_2 layers.

1. INTRODUCTION

A striking feature of $YBa_2Cu_3O_7$ is the sensitivity of its superconducting, electrical, magnetic ... properties to changes in its oxygen content. Since the discovery of high-temperature superconductivity in this compound, detailed neutron-diffraction investigations of the crystal structure of $YBa_2Cu_3O_x$ have revealed that the distribution of the oxygen ions among the various crystallographic O sites of the triple-perovskite unit cell[1] is partly disordered when the oxygen stoichiometry x deviates from 7.[2] The most important structural effect set up by reducing the oxygen content x is a continuous depopulation of the O(1) sites that bond the O(4)-Cu(1)-O(4) units,[3] oriented parallel to the c axis, along one-dimensional ribbons parallel to the b (>a) axis. Simultaneously, the O(5) sites between the Cu(1)-O(1) chains, from empty in the x=7 compound, are progressively but partially occupied. Eventually, the system undergoes an orthorhombic — tetragonal transition[4] when the occupancies of the adjacent O(1) and O(5) sites in the mirror plane of the unit cell reach a common value close to 0.25.[2,5]

Complementary electron-diffraction studies indicate that the oxygen vacancies in $YBa_2Cu_3O_x$ are not always distributed at random among the O(1) and O(5) sites for x

≈ 6.5, but often form ordered superstructures.[6,7] In possible connection with structural modifications, changing the oxygen content in $YBa_2Cu_3O_x$ affects the superconducting transition temperature T_c, which ranges from above 90 K for $x \approx 7.0$ down to about 20 K at $x \approx 6.4$, with a plateau ($T_c \approx 60$ K) for $6.5 < x < 6.7$.[8] In the tetragonal phase, $YBa_2Cu_3O_x$ is no more a superconductor and becomes an antiferromagnetic insulator when x approaches 6.9[9] The antiferromagnetic ordering of the Cu moments in the "undoped" $x \approx 6$ material takes place in the (a,b) square-planar CuO_2 networks[9] and, presumably, the superconducting properties of the YBaCuO system also have their origin in the CuO_2 planes, at least most of the theories of high-T_c superconductivity proposed so far agree on that point. However, reducing the oxygen content x in $YBa_2Cu_3O_x$ mainly disrupts the Cu-O chains without affecting significantly the CuO_2 layers.[10] Consequently, the sensitivity of the properties of the YBaCuO system to changes in its oxygen stoichiometry implies some coupling between the CuO_2 planes and the Cu-O chains, and the origin of this coupling is not yet well understood.

In this paper, the effects of both the oxygen content and oxygen (dis)ordering on the electronic structure of $YBa_2Cu_3O_x$ in the normal state are investigated within a tight-binding approximation. The electronic density of states of the recently synthesized $YBa_2Cu_4O_8$ 80 K superconductor[11] is also computed. In addition, the electronic properties of an idealized (110) twin wall in $YBa_2Cu_3O_7$ are analyzed. The recursion algorithm that has been used allows us to treat random oxygen occupancies with the same degree of accuracy as ordered configurations, as this real-space technique does not appeal to Bloch's theorem.[12] A special emphasis is put on two band-structure properties which may be of some interest as far as superconductivity is concerned : the density of states $N(E_F)$ at the Fermi energy on the one hand and, on the other hand, the number of unoccupied states in the electronic antibonding bands which have their origins in the CuO_2 layers and cross the Fermi level.[13,14,15] If the usual Cooper-pairing of electrons (or holes) prevailed in some neighborhood of the Fermi level, a crucial quantity to examine would be $N(E_F)$ — more precisely its value averaged over an interval spanning the energy spectrum of the relevant pairing bosons — and, indeed, some correlations between measured $T_c(x)$ and $N(E_F)$ values computed in $YBa_2Cu_3O_x$ have already been emphasized.[15,16] On the other hand, according to most of the current high-T_c theories, the antiferromagnetic / superconducting phase diagram of both the LaCuO and YBaCuO systems is believed traducing frustration effects in the $Cu(3d^9)$ spin ordering as the number of Cu(3d)-O(2p) holes in the CuO_2 sheets, from one per formula unit in the antiferromagnetic compound, raises by doping the structure with electron acceptors. In fact the $T_c(x)$ behavior in $YBa_2Cu_3O_x$ has been interpreted in terms of the dependence on x of the numbers of holes in the CuO_2 layers, as determined by electronic-structure calculations assuming O vacancies distributed regularly on the O(1) sites.[17] These two approaches, exploring the variations of, respectively, $N(E_F)$ and the hole count in the CuO_2 planar networks with changes in the oxygen concentration and/or the O-vacancy ordering, are reconsidered below.

2. THE TIGHT-BINDING HAMILTONIAN

The present paper is based on a tight-binding description of the YBaCuO electron structure that has been adapted from a model proposed by Richert and Allen.[18] After a brief description of this model, we proceed in simplifying the orbital basis set so as to optimize the model for its applications to the recursion method.

Table 1. Tight-binding parameters for the YBaCuO system (adapted from Ref. 18).

	ε_s (eV)	ε_p (eV)	ε_d (eV)	r_d (Å)	Strengths (eV) of the two-center integrals
O	-29.0	-14.0			$\eta_{sd\sigma} = -1.6$ $\eta_{pd\sigma} = -2.5$ $\eta_{pd\pi} = 1.4$
Cu	-11.0		-14.0	0.95	$\eta_{ss\sigma} = -1.1$ $\eta_{sp\sigma} = 0.9$
Y	-5.3		-6.8	1.6	
Ba	-4.5		-5.5	1.8	

2.1. Semi-empirical parameters

The semi-empirical tight-binding description of the electronic structure of the cuprate superconductors originally developed by Richert and Allen include s and p orbitals for the oxygen atoms and s and d orbitals for Cu, La, Ba or Y.[18] Only first-neighbor hopping interactions between O and the metal elements M are taken into account. The interatomic M-O matrix elements are computed within the two-center approximation, using Harrison's semi-empirical expressions[19] for the dependence of the two-center integrals on the interatomic distances d :

$$V_{ll'm} = \eta_{ll'm} \frac{\hbar^2}{md^2} \;(1, 1' = s \text{ or } p) \;;\; V_{ldm} = \eta_{ldm} \frac{\hbar^2 r_d^{3/2}}{md^{7/2}} \;(1 = s \text{ or } p) . \quad (2.1)$$

The strengths $\eta_{ll'm}$ and the size r_d of the d orbitals in eqs. (2.1), together with the orbital energies ε_l of the tight-binding Hamiltonian, which were adjusted to the band structure of La_2CuO_4, were transposed without further changes to other HiTc oxides,[18] including the Bi- and Tl-based compounds.[20] For applications to $YBa_2Cu_3O_7$, however, we find better agreement with *ab-initio* band structures[13-15] by modifying some of the orbital energies proposed by Richert and Allen (see Table 1 for a list of the parameters we have employed in our calculations). In so doing, we recover a ~0.6 eV energy gap around $E_F + 2eV$, absent in the original tight-binding calculations of Ref. 18, which separates the Cu(3d)-O(2p) manifold of bands from the empty "conduction bands" with Cu(4s), Ba(5d) and Y(4d) characters.[13-15]

2.2. Reduction of the orbital basis set

Since our main interest will be focused on the Cu(3d)-O(2p) manifold of bands that comprises the Fermi level, we now proceed in simplifying the orbital basis set. In particular, our main objective is to discard the occupied O(2s) states that form narrow bands 10 eV below the bottom of the Cu(3d)-O(2p) manifold. Our motivation in removing the O(2s) orbitals is to disentangle the band structure from this wide 10 eV gap, because the recursion technique we want to apply is not well adapted to deal with a gap of that size, comparable with the width of the allowed energy bands. By the same way, the Ba(6s) and Y(5s) orbitals have also been removed from the basis set, as these orbitals give rise to non-occupied states well above the Fermi level.

The reduction of the orbital set has been performed using a Löwdin correction of the tight-binding Hamiltonian H_{AA} related to the subset A of the orbitals retained.[21] Mathematically, this correction technique can be understood as follows. The secular equation for the full matrix Hamiltonian H is contracted according to the well-known identity

$$
\text{dtm} \begin{vmatrix} H_{AA}\text{-}E & H_{AB} \\ H_{BA} & H_{BB}\text{-}E \end{vmatrix} = \text{dtm} \begin{vmatrix} H_{BB}\text{-}E \end{vmatrix} \cdot \text{dtm} \begin{vmatrix} H_{AA}\text{-}E \text{ - } H_{AB}(H_{BB}\text{-}E)^{-1}H_{BA} \end{vmatrix}
$$

(2.2)

provided E be outside the energy spectrum of the Hamiltonian H_{BB} related to the subset B of the orbitals removed. Within the energy range of interest, eq. (2.2) can further be approximated to an eigenvalue problem in the subspace A, considering the "Löwdin-corrected" matrix

$$
\tilde{H}_{AA} = H_{AA} \text{ - } H_{AB}(H_{BB}\text{-}\overline{E})^{-1}H_{BA}
$$

(2.3)

where $\overline{E}$ denotes some mean energy within the range of interest. We chose $\overline{E}$ = -12 eV with respect to the zero of energy relevant to the one-site energies given in Table 1, that is to say close to the Fermi level of $YBa_2Cu_3O_7$ that was found at about -11.5 eV. Besides, eq. (2.3) can further been simplified by retaining only the sole diagonal part of the matrix H_{BB}. This approximation is fully justified in the present context as the off-diagonal elements of this matrix couple Y(5s) or Ba(6s) to O(2s) orbitals, the energy separations between the orbitals involved (~24 eV, see Table 1) being much larger than the interatomic coupling elements (~1.5 eV for the strongest Y(5s)-O(2s) bond).

Physically, the correction of the matrix H_{AA} arising from eq. (2.3) has two effects. First it modifies the diagonal elements by introducing some sort of a crystal-field splitting of the levels of the orbital retained : we find a ~0.2 eV and ~0.4 eV splitting of the Cu(3d) and O(2p) levels, respectively, with the average O(2p) level ~0.3 eV below the average Cu(3d) on-site energy. Also the Cu(4s) levels are pushed upwards by ~1.3 eV. Next the Löwdin correction introduces O-O and M-M interactions (M denoting a metal element), otherwise absent in the full Hamiltonian. Amongst these new interactions, the O-O interatomic elements were found the largest and, for that reason, only these O-O elements have been taken into account in addition to the first-neighbor M-O interactions already included in H_{AA}. As an example, we obtain a matrix element $E_{x2,y3}(1/2,1/2,0) = 0.17$ eV coupling the p_x and p_y orbitals on adjacent O(2) and O(3) sites in the CuO_2 sheets. This interaction, here promoted by the neighbor Y and Ba_2 s-orbitals through the Löwdin correction (eq. 2.3), has a value that compares well with that deduced by fitting the LAPW band structure of "classical" perovskites like $SrTiO_3$.[22]

The Hamiltonian (eq. 2.3) contracted in the reduced orbital basis set is our starting point for investigating the effects of ordered and disordered defects in the YBaCuO system. The band structure of $YBa_2Cu_3O_7$ obtained is shown in Fig. 1. The structure of the manifold of the 36 bands with Cu(3d)-O(2p) character has been found almost indistinguishable from that derived from the full tight-binding Hamiltonian. Although the essential features of the *ab-initio* band structure of $YBa_2Cu_3O_7$ are correctly reproduced in Fig. 1, fine details are out of reach of the present tight-binding description. For instance, the nearly flat chain band at 1.5 above E_F has a (too) small dispersion along the MY symmetry line, leading therefore to a peak in the density of states (see e.g. Fig. 4 below) bearing strong resemblance with a one-dimensional band-edge singularity. Also we do not obtain a fourth, chain-related band crossing the Fermi level, while remaining close to E_F all along the MY and YΓ lines.[23] For all these reasons, the Fermi level is pushed too high by about 0.2 eV and the density of states at the Fermi energy deduced from the band structure of Fig. 1 is too small, roughly by a factor 2, with respect to LAPW[13] or pseudofunction[15] values, while being more consistent with the LMTO calculations of Ref. 14.

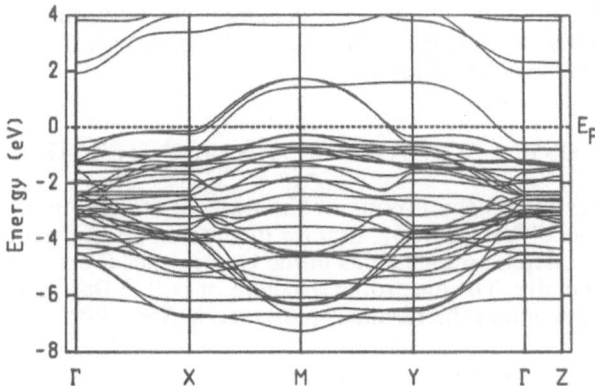

Fig. 1.   Tight-binding band structure of $YBa_2Cu_3O_7$ obtained in a reduced basis of 54 orbitals per unit cell, as explained in the text.

2.3. The recursion technique

The recursion method enables expanding diagonal elements of the resolvent operator $(z-H)^{-1}$ in a continued-fraction form

$$R_{I\lambda}(z) \equiv \langle I\lambda| (z-H)^{-1} |\lambda I\rangle = \cfrac{1}{z-a_1 - \cfrac{b_1^2}{z-a_2 - \dots}} \tag{2.4}$$

from algebraic manipulations of the matrix Hamiltonian in direct space.[12] In the above expression, λ denotes a particular orbital s, p_x, ..., $d_{3z^2-r^2}$ located on a given atomic site I of the crystal. The local density of states $n_I(E)$ projected on this site follows from the usual relation $n_I(E) = -1/\pi \sum_\lambda R_{I\lambda}(E+i0)$.

For three dimensional systems, the memory storage on the computer and the cpu time for obtaining 2n recursion coefficients a_1, b_1 ... a_n, b_n increases like n^3 and, therefore, only a small number of coefficients can be computed in practice. In this paper, n=10 pairs of coefficients have been computed on clusters of, typically, 10×10×4 YBaCuO unit cells (5200 atoms), with periodic boundary conditions applied in the three directions. The recursion coefficients have been extrapolated to higher levels until constant-coefficient terminations of the continued-fraction expressions (eq. 2.4) can be applied.[24] The densities of states so determined are only exact to within a "resolution" roughly equal to the total bandwidth — that has been considerably reduced by removing the O(2s) occupied bands — divided by the number 2n of exact recursion coefficients. For instance, the densities of states shown below assume small, but non zero, values in the gap at about 2 eV above the Fermi level (see Fig. 1). Nevertheless, in view of the uncertainty affecting the parameters of Table 1, we consider that the resolution obtained from only ten pairs of recursion coefficients is sufficient for a qualitative description of the effects of the oxygen stoichiometry on the electronic properties of the YBaCuO system. We have checked that the essential features of the tight-binding density of states of $YBa_2Cu_3O_7$ computed from the band structure by a conventional Brillouin-zone integration technique are correctly reproduced when obtained by the recursion technique, especially in the vicinity of the Fermi level where, indeed, a good agreement was observed.

3. BULK ELECTRONIC STRUCTURE OF THE YBaCuO SYSTEM

We have investigated the effects of the oxygen stoichiometry x on the bulk electronic properties of $YBa_2Cu_3O_x$, first for disordered O configurations (Sect. 3.1) and next for some ordered arrangements of the O vacancies (Sect. 3.2) that have been observed experimentally. The electronic structure of bulk $YBa_2Cu_4O_8$ is analyzed in Sect. 3.3 and a discussion of the results is presented in Sect. 3.4.

3.1. Disordered oxygen vacancies

Throughout this paper, the O vacancies generated by reducing the oxygen content in $YBa_2Cu_3O_x$ are supposed to disrupt the Cu-O chains only. In so doing, we neglect the small depopulation of the O(4) (BaO-sheet related) sites observed in samples with stoichiometries smaller than 7,[10] a consequence of which being that the orthorhombic — tetragonal transition is actually realized for a total oxygen content $x \approx 6.4$ whereas only about 0.5 O-sites in the chains have been depleted. Denoting by δ the fraction of O-sites depleted in the Cu(1)-O(1) (chain) mirror plane when reducing the oxygen concentration in the YBaCuO system, tight-binding densities of states of $YBa_2Cu_3O_{7-\delta}$ have been computed by assuming uncorrelated and random occupancies x_1 and x_5 of the O(1) (chain bonds) and O(5) (between the chains) sites. We have adopted the following site-occupancy probabilities

$$x_5 = \frac{1 - \sqrt{1-(2\delta)^2}}{4} \quad (\delta \leq 0.5) \ , \quad x_5 = \frac{1-\delta}{2} \quad (0.5 \leq \delta \leq 1.0)$$

$$x_1 = 1 - \delta - x_5 \tag{3.1}$$

which well reproduce experimental data for the high-temperature related structures.[2,10] The above expressions place the boundary between the orthorhombic ($x_1 > x_5$) and the tetragonal ($x_1 = x_5$) phases at $\delta = 0.5$, where the long-range order parameter $x_1 - x_5$ vanishes like $(1-2\delta)^{1/2}$; for small values of δ, x_1 from one in $YBa_2Cu_3O_7$ ($\delta = 0$), decreases linearly with δ whereas the O(5) sites remains almost empty.

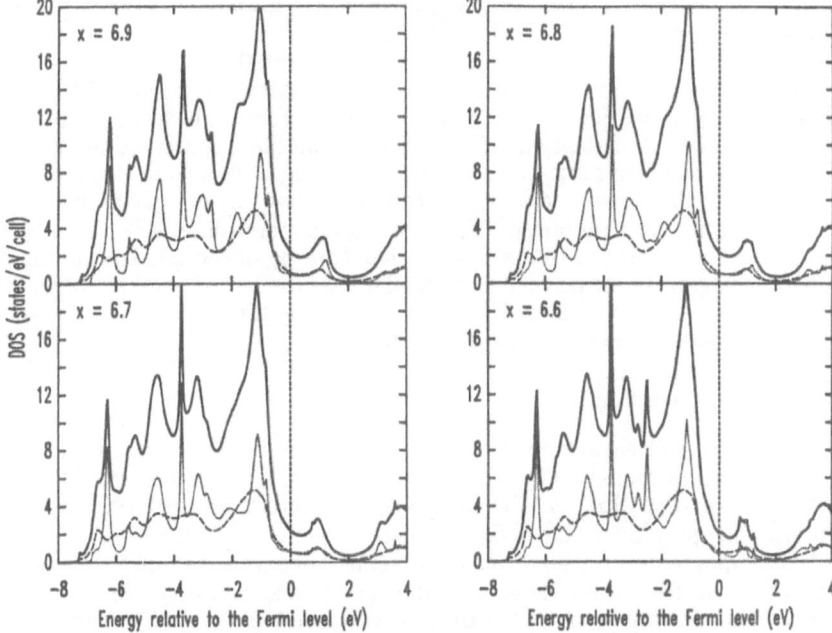

Fig. 2.   Configurational-averaged densities of states of $YBa_2Cu_3O_x$ for x = 6.9 to 6.6 for random occupancies of the O(1) and O(5) crystallographic sites given by eqs. (3.1), modelling the high-temperature related structures. The dotted curves visualize the distribution of states arising from the (disrupted) Cu-O ribbons (composed of the Cu(1), 2×O(4), O(1) and O(5) sites) ; the contribution per formula unit of the CuO_2 layers are indicated by the dashed curves.

Variations of the lattice parameters *vs* the oxygen stoichiometry have been taken into account, simply by scaling the crystallographic structure of the δ=0 compound so as to match the actual lattice parameters while neglecting relaxations of the atomic positions within the unit cell.[10] Tight-binding calculations were then performed following a procedure already applied to disordered transition-metal alloys.[25] In a given unit cell of a 10×10×4-cell cluster, where partial densities of states have been computed, we have taken into account the four possible O configurations resulting from the occupation or the non occupation of the O(1) and O(5) sites. For each of these four O distributions in the central cell, those oxygen or vacancy configurations of the neighboring cells that have high statistical weights have been generated, whereas the occupation or non-occupation of the O(1) and O(5) sites in the rest of the cluster have been taken at random, according to the occupancy probabilities given by eqs. (3.1). The configurations so generated have been weighted and the densities of states have been averaged over these configurations (whose number ranged from 56 for δ = 0.1 to 160 for δ = 0.5).

Fig. 2 shows the configurational average of the density of states of $YBa_2Cu_3O_{7-\delta}$ for δ = 0.1 - 0.4. One first notices that the oxygen stoichiometry does not affect the distribution of the electron states arising from the CuO_2 layers (dashed curves).

Obviously, the electronic structure of the ribbons (dotted curves) is much more sensitive to the oxygen content, particularly in some ranges of energy where the ribbon states are weakly hybridized with those arising from the CuO_2 planes. So is for instance the region around 1 eV above E_F, and to a lesser extend that one spanning a small interval around the Fermi level, where the progressive appearance of peaks in the ribbon states traduces localization effects arising from the fragmentation of the chains into isolated islands when more and more binding oxygen atoms are removed from the structure. In this connection, small chain fragments have been proved leading to flat bands in the electronic structure of $YBa_2Cu_3O_{6.5}$ superstructures[26] ; locally, strong enhancements of the density of states may therefore be observed. Smoothing effects due to disordering occur, however, in particular near the Fermi level as shown in Figs. 2 and 3. The peak at 3.7 eV below E_F is an O(4) state getting more and more localized character as oxygen concentration x decreases.

3.2. Ordered (super)structures

Ordering of the O vacancies in $YBa_2Cu_3O_x$ at some intermediate stoichiometries has been observed or predicted by a number of authors. For instance, doubling the unit cell along the a direction in $YBa_2Cu_3O_{6.5}$ leads to an ordered ground-state configuration keeping every other Cu-O chains intact ($x_1 = 1/2$, $x_5 = 0$).[27] Signatures of such a superstructure have indeed been observed in electron-diffraction experiments.[6,7] In Fig. 3, the density of states of ($2a \times a \times c$) $YBa_2Cu_3O_{6.5}$ presenting such an ordered arrangement of the oxygen vacancies is compared with that of the tetragonal ($a \times a \times c$) phase having its O(1) and O(5) sites filled at random with $x_1 = x_5 = 0.25$. Here again, the electronic structure of the CuO_2 layers (dashed curves) is only weakly affected by the particular O-vacancy arrangement generated in the Cu(1)-O(1)-O(5) mirror plane. Referring to the density of states of the ordered structure, the small peak in the dotted curve (ribbon contribution) at 1.3 eV above E_F is the band-edge singularity associated with the intact Cu-O chains ; the peak at $E_F + 3$ eV is due to the isolated O(4)-Cu(1)-O(4) "molecules" left by removing all the O(1) bonding atoms in every other chains. As for the disordered case, the multi-peak structure around 1 eV above E_F traduces localization effects already discussed in Sect. 3.1. The electronic contribution to the internal energy (at 0 K) of the ordered superstructure has been found 0.15 eV/cell below that related to the random O distribution.

We have also analyzed the electronic properties of the ordered ($2\sqrt{2}a \times 2\sqrt{2}a \times c$) superstructure that has been inferred from electron-diffraction data in some samples of $YBa_2Cu_3O_{7-\delta}$ for δ close to 1/8.[28] Here, every fourth O(1) atom is removed from every other Cu(1)-O(1) chain in a periodic pattern, all the O(5) sites remaining empty. The density of states obtained for this superstructure has been found very close to that computed for a random configuration of the O vacancies, using site occupancies given by eqs. (3.1) for the same $\delta = 1/8$ composition (see Table 2). Calculations show that the electronic contributions to the internal energies of both the ordered superstructure and the disordered phase are virtually identical.

3.3. The "Y124" phase

Recently, the $YBa_2Cu_4O_8$ 80 K superconductor has been synthesized under high oxygen pressure.[11] Neutron diffraction data have revealed the crystallographic structure of this one-face, A-centered orthorhombic compound : it results from the intercalation of a supplementary Cu(1)-O(1) (chain sites) plane in the "123" unit cell, with a b/2 shift of half a cell.[29] Similar double Cu-O layers between the two BaO sheets, associated with a 15 % increase of the c axis, were already observed as stacking faults in $YBa_2Cu_3O_7$ by high-resolution transmission electron microscopy.[30] The tight-binding density of states of the "124" compound is compared with that of the usual "123" material in Fig. 4 . It is worth noticing that the CuO_2 sheets have mainly identical electron-state distributions (dashed curves) in the two compounds. As for the tight-binding band structure (not illustrated) of $YBa_2Cu_4O_8$, it differentiates from that of $YBa_2Cu_3O_7$ in this that there are now four bands crossing the Fermi level : two bands originate from the two CuO_2 sheets as before (Fig. 1), the other two bands (originally one) arising from the double-chain structure.

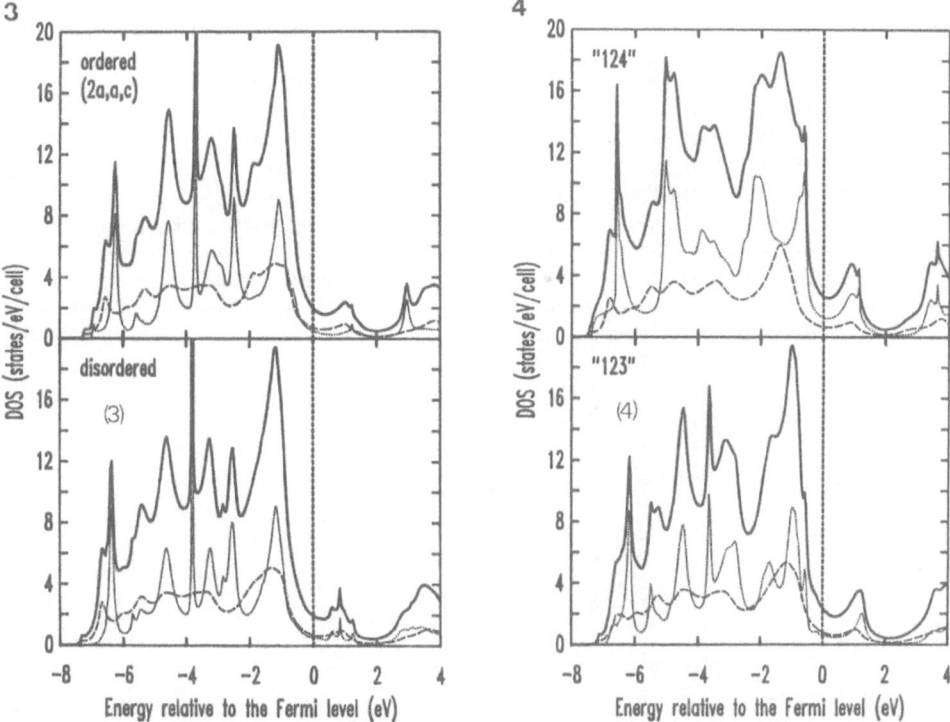

Fig. 3. (left) Density of states of $YBa_2Cu_3O_{6.5}$ for two configurations of the oxygen sites in the mirror plane of the unit cell. In the ordered superstructure (top), the O(1) sites along the Cu-O chains are alternatively occupied and vacant with period 2a, whereas all the O(5) sites between the chains are unoccupied. By contrast, these two sites are equally populated in the disordered, tetragonal phase (bottom), with an occupancy fraction equal to 0.25.

Fig. 4. (right) Tight-binding densities of states of the "123" $YBa_2Cu_3O_7$ compound (bottom) and the "124" $YBa_2Cu_4O_8$ material (top). In both of these figures, the dashed and dotted curves visualize the distributions of the electronic states arising from the CuO_2 layers and the ribbon structures, respectively.

Table 2. Electronic properties of bulk $YBa_2Cu_3O_{7-\delta}$ and $YBa_2Cu_4O_8$. The occupancy fractions of the O(1) and O(5) sites are denoted by x_1 and x_5, respectively, as given by eqs. 3.1 for those random configurations (R) modelling the high-temperature related structures ; the ordered superstructures (O) are described in the text. ΔE_F : shift of the Fermi level with respect to that in $YBa_2Cu_3O_7$; $N(E_F)$: density of states at the Fermi energy ; p_{CuO_2} : hole number per formula unit in the CuO_2-layer distribution of states.

Composition			ΔE_F (eV)	$N(E_F)$ (st./eV/cell)	p_{CuO_2} (hole/f.u.)	Structure	
δ	x_1	x_5					
0.0	1.00	0.00	0.00	2.42	1.25	conventional "123" structure	O
0.1	0.89	0.01	0.04	2.32	1.21	orthorhombic	R
1/8	0.87	0.01	0.06	2.32	1.19	orthorhombic	R
1/8	7/8	0.00	0.06	2.30	1.20	$(2\sqrt{2}a \times 2\sqrt{2}a \times c)$	O
0.2	0.78	0.02	0.08	2.25	1.17	orthorhombic	R
0.3	0.65	0.05	0.12	2.26	1.13	orthorhombic	R
1/3	2/3	0.00	0.06	2.24	1.17	$(3a \times b \times c)$	O
0.4	0.50	0.10	0.14	2.07	1.10	orthorhombic	R
0.5	0.25	0.25	0.19	1.93	1.06	tetragonal	R
0.5	0.50	0.00	0.08	2.04	1.14	$(2a \times a \times c)$	O
1.0	0.00	0.00	0.20	1.29	0.98	non-magnetic, tetragonal	O
$YBa_2Cu_4O_8$			0.25	2.69	1.08	orthorhombic	O

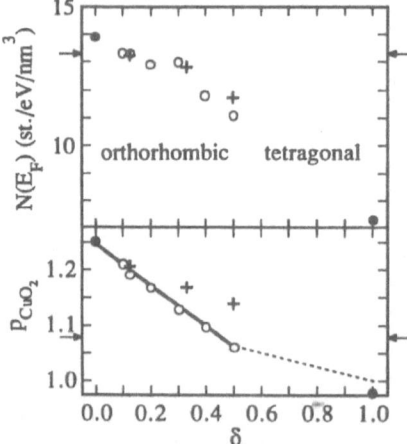

Fig. 5.   Density of states per <u>unit volume</u> at the Fermi energy (top) and hole number per formula unit in the CuO_2 layers (bottom) in $YBa_2Cu_3O_{7-\delta}$. The filled circles are related to $YBa_2Cu_3O_7$ and $YBa_2Cu_3O_6$. The open circles correspond to high-temperature related structures modelled by randomly occupied O(1) and O(5) sites, with occupancy fractions computed by eqs. (3.1). The cross markers are related to the ordered superstructures listed in Table 2. The arrows indicate the values of $N(E_F)$ and p_{CuO_2} realized in $YBa_2Cu_4O_8$.

3.4. Discussion

Electronic parameters of bulk $YBa_2Cu_3O_{7-\delta}$ and $YBa_2Cu_4O_8$ are listed in Table 2. Restricting first our attention to $YBa_2Cu_3O_x$, one notices that removing oxygen from $YBa_2Cu_3O_7$ shifts the Fermi level upwards : the oxygen vacancies act as donors of electrons, as already observed by photoemission spectroscopy[31] and predicted from *ab-initio* band-structure calculations for the x=8, 7 and 6 compounds.[14,15,32] Although a coherent-potential approximation leads to an opposite behavior in La_2CuO_{4-y},[33] the upward shift of E_F observed in $YBa_2Cu_3O_7$ on removing O is easily understood by noticing that each oxygen accounts for six p-electron states in the band structure but contributes four p-electrons to the band filling.

The behavior of the density of states $N(E_F)$ at the Fermi energy of $YBa_2Cu_3O_{7-\delta}$ is plotted in Fig. 5 (top) against δ. These data essentially traduces variations of the density of states at the Fermi level originating from the ribbon structure on going from $\delta = 0$ to $\delta = 1$. The ribbons contribute nothing to $N(E_F)$ in $YBa_2Cu_3O_6$: the O(4)-Cu(1)-O(4) units left in the latter structure after removal of all the chain-binding O(1) atoms[34] are found insulating from 0.3 eV below the Fermi level to 2.5 eV above E_F ; the total density of states at E_F listed in Table 2 drops approximately by a factor 2 when δ ranges from 0 to 1, as already pointed out e.g. in Ref. 14. Let us emphasize that the recursion technique allowed us "interpolating" between the two limiting x=7 and x=6 "normal-state" (i.e. non superconducting or non-magnetic) compounds while going beyond the virtual-crystal approximation or even the coherent-potential approximation as far as disordering effects are concerned. Though the tight-binding absolute values of $N(E_F)$ for these two limiting x=6 and x=7 compounds might be in error by a factor as large as two, a general trend is that $N(E_F)$ is a decreasing function of δ (apart from a local enhancement around $\delta \approx 0.3$), in agreement with a monotonous decrease of $N(E_F)$ already predicted by a Green's function technique applied to independent O vacancies in $YBa_2Cu_3O_x$.[16] One also notices that the density of states at E_F of the ordered super-structures analyzed in Sect. 3.2 do not differentiate very much from that realized in the high-temperature related structures having random occupancies of their O(1) and O(5) sites, considering that fine details of the tight-binding density of states are washed out by the limited accuracy of the recursion technique. In addition the density of states at the Fermi level of the "Y124" 80 K superconductor, indicated by arrows in Fig. 5, fits well that realized in $YBa_2Cu_3O_{\sim6.85}$, these two compounds having similar T_c's.

Since the electron states arising from the CuO_2 sheets are merely insensitive to the oxygen distribution in the Cu(1)-O(1) (chain) mirror plane, the displacement of the Fermi level relative to the "two-dimensional" bands on going from x = 7 to x = 6 has an important consequence on the hole number in the CuO_2-layer electron states (i.e. the number of unoccupied states in the Cu(3d)-O(2p) antibonding "two-dimensional" bands). This consequence is clearly shown in Fig. 5 (bottom) in the form of a lowering of the hole count p_{CuO_2} when δ ranges from 0.0 to 0.5 in $YBa_2Cu_3O_{7-\delta}$. The hole number p_{CuO_2} has been computed by integrating the CuO_2 electron density of states (shown by the dashed curves in Figs. 2-4) from E_F to the bottom of the Cu(3d)-O(2p) manifold of bands.[35] (By integrating the CuO_2-projected density of states, the small Ba or Y character of the CuO_2 bands arising from hybridization is neglected : for that reason, p_{CuO_2} is found equal to 0.98 in $YBa_2Cu_3O_6$ instead of being exactly one.) We do not observe any special behavior of p_{CuO_2} that could be attributed to chain fragmentation effects such as reported in Ref. 17, where O-vacancy distributions different from ours have been postulated (but we do observe such effects on the density of states, as pointed out above). On the contrary, we find a monotonous variation of the hole number vs δ,[36] with a slope $dp/d\delta \approx -0.4$, for those random occupations of the O(1) and O(5) sites investigated in Sect. 3.1. This behavior might be correlated with the continuous lowering of T_c with decreasing oxygen stoichiometry observed in some quenched materials.[2,37] As already mentioned in the Introduction, indeed, it is tempting in view of current HiTc theories to correlate T_c with the hole count in the two-dimensional CuO_2 bands. Given that the tetragonal phase of $YBa_2Cu_3O_x$ is not superconducting, one could hypothesize that superconductivity is destroyed when the hole count in the CuO_2 sheets falls below a certain threshold that may be estimated at $p_c \approx 1.05$ from the data of Table 2. Such a value is also consistent with data taken from $La_{2-y}Sr_yCuO_4$,[38] whose superconducting phase covers a doping concentration domain above $y \approx 0.05$ (assuming $p_{CuO2} \approx 1+y$ in this material).

The ordered double-cell superstructure observed[6] in $YBa_2Cu_3O_{6.5}$ is characterized by a CuO_2-layer hole number significantly higher than that of the disordered, tetragonal compound having the same composition (see Fig. 5). Intimately connected with the stabilization of this ordered orthorhombic phase (OII) at stoichiometry x = 6.5, Monte-Carlo simulations of O ordering in the mirror plane of $YBa_2Cu_3O_{7-\delta}$ indicate[39] that metastable ordered structures having fully-formed Cu-O chains may be realized at low temperature for $\delta \neq 0.5$; more specifically, groups composed of a variable number of adjacent intact chain(s) in the mirror plane of the structure are found never separated by more then one empty chain when $\delta < 0.5$. To further investigate the effects that fully-formed chains have on the electronic properties of the YBaCuO system, we have considered an ordered triple-cell superstructure, locally present in the Monte-Carlo simulations[39] and not inconsistent with electron diffraction data,[7] consisting of alternating full-full-empty chains as an idealization of the $\delta=1/3$ configuration. Tight-binding calculations performed for this ($3a \times b \times c$) triple cell undoubtedly prove that superordering of the oxygen vacancies is responsible for increasing the hole number p_{CuO2}, as already found for the OII phase of $YBa_2Cu_3O_{6.5}$. Enhanced hole doping brought about by intact-chain configurations is clearly illustrated in the bottom part of Fig. 5 by the crosses (+) at $\delta = 1/3$ and 1/2. As suggested by a number of authors, therefore, the ~60 K plateau observed under special preparation conditions in the $T_c(x)$ curve of $YBa_2Cu_3O_x$ around x $\approx$ 6.6 can be attributed to superordering effects that involve alternating intact and empty Cu-O chains, raising the CuO_2 hole concentration, as illustrated by Fig. 5, and the superconducting transition temperature as well if one believes that a correlation exists between these two quantities.[17] However, the hole number we have obtained for the $YBa_2Cu_4O_8$ 80 K superconductor (indicated by the arrows in Fig. 5) does not support a straightforward correlation between T_c and the hole number in the CuO_2 sheets ($T_c \sim p_{CuO2}-p_c$, say). If our figures are correct, indeed, the hole count (1.08) in $YBa_2Cu_4O_8$ will place this superconductor just above the threshold p_c defined here above and, in these conditions, it will be difficult to understand solely from hole counting how a critical temperature T_c as high as 80 K can be achieved in this material. Of course, the question of the validity of our tight-binding Hamiltonian, and/or its parametrization, can be raised. Clearly, further investigations from "state of the art" band-structure calculations are called for in this context. However, we can argue that hole numbers, being an integral property of the density of states, should not be too sensitive to fine details of the band structure.

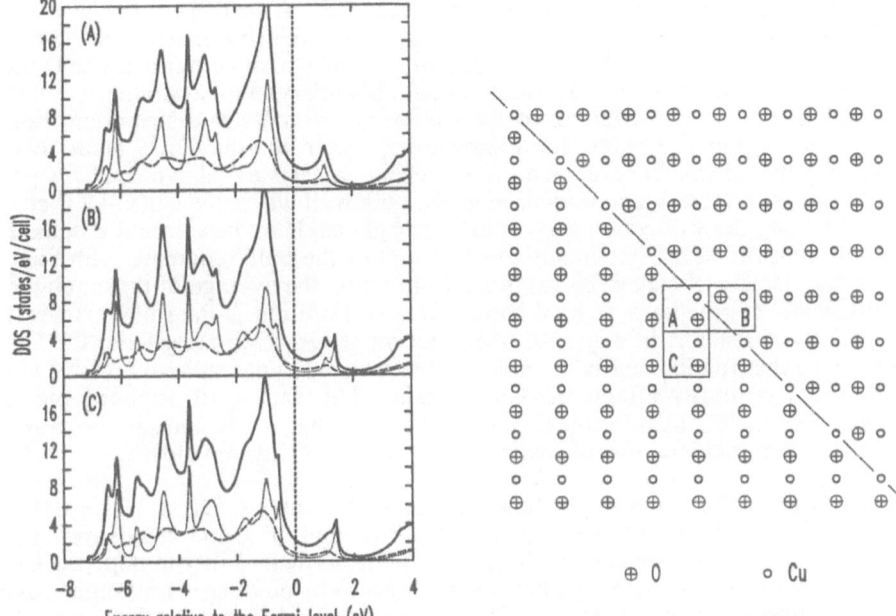

Fig. 6. Density of states of twinned $YBa_2Cu_3O_7$ computed in the unit cells marked
A, B and C in the drawing of the right-hand side, where only the Cu-O chains
have been represented, for the case of an idealized (110) twin wall. The
dotted and dashed curves in the left-hand side of the figure have the same
meanings as in Figs. 2-4.

4. ELECTRONIC PROPERTIES OF TWINNED $YBa_2Cu_3O_7$

Detailed electron microscopy experiments have revealed several structural defects
in the YBaCuO system.[30,40] Among the planar defects inventoried,[41] stacking faults
from which the "Y124" structure originates have already been mentioned in Sect. 3.3.
In this section, we turn our attention to twin boundaries. It is a well-known fact that
orthorhombic $YBa_2Cu_3O_{7-\delta}$ is heavily twinned.[42] Although it is likely that twinning
plays a minor role, if any, in the mechanisms responsible for superconductivity in these
materials, it is not without interest to analyze their effects on the electronic structure of
$YBa_2Cu_3O_7$. Considering the case of (110) mirror twins, in which the a and b axes are
interchanged across the twin boundary, we have adopted the simple atomic
arrangement consisting in Cu(1)-O(1) chains rotated by 90°-ε right at the twin wall
(Fig. 6).[43] The twin obliquity ε, being small (~1°), has been discarded in our
calculations, this simplification consisting in neglecting the small difference between
the a and b crystallographic parameters.

The left-hand side of Fig. 6 shows the densities of states of twinned $YBa_2Cu_3O_7$
that have computed in the three cells indicated on the right-hand side. Each of these

cells contains the 13 atoms of the "123" formula unit. With respect to the perfect, untwinned crystal, there is a slight enhancement of the density of states at the Fermi level right at the twin wall (see Fig. 6 A). In cell A, which contains a Cu(1) atom (shared by two orthogonal half chains at the twin boundary) with asymmetric fourfold coordination, there is a flattening of the ribbon electron states distribution (dotted curve) in the region just above the Fermi level, much like the CuO_2-layer related density of states (dashed curve), with however a peak at 1 eV above E_F. This peak, with O(1) p_y character, is a consequence of the twin wall where the Cu(1)-O(1) chains are broken along the y direction and continue at right angle in the adjacent B cells. The density of states in cell C is already close to that of the bulk (compare with Fig. 4), whereas the density of states (B) is not far from being the average of those related to the cells A and C ; the peak at 1 eV above E_F is still present in the ribbon states (B), with O(1) p_x character. In any case, the electron states originating from the CuO_2 sheets have a distribution nearly identical to that realized in the bulk, insensitive to the twinning that primarily affects the Cu-O chains. This fact again supports the idea originally developed by Mattheiss and Hamann,[13] that the electronic structure of YBaCuO system is chiefly that of weakly-interacting Cu-O layered blocks.

In the three cells of Fig. 6, the hole number in the CuO_2 sheets is 1.25 ± 0.01 per formula unit, just like in the bulk material. Even though the Cu-O chains form a $90°$ angle at the twin wall, they behave like intact chains in their ability to dope the CuO_2 layers by holes. Let us finally emphasize that our tight-binding Hamiltonian favors such a twinning, as the one-electron energy of the twinned material is lowered by the formation across the (110) boundary of a new O-O bond (mediated by the s-orbitals of neighboring Ba atoms shared by the two O(1) atoms in cells (A) and (B), as explained in Sect. 2.2). The electronic contribution to the interfacial energy of the twin boundary we have found is $\gamma_{twin} \approx -0.09$ J/m^2.

ACKNOWLEDGEMENTS

This work has been supported by the Concerted Action Program IRIS (Institut de Recherche sur les Interface Solides). The use of the Namur Scientific Computing Facility, a common project between the Belgian National Foundation for Scientific Research, IBM-Belgium and the Facultés Universitaires Notre-Dame de la Paix, is greatly acknowledged.

REFERENCES

*   Research Associate at the Belgian National Foundation for Scientific Research.
1.   J.E. Greedan, A.H. O'Reilly, and C.V. Stager, Phys. Rev. B 35:8770 (1987) ;
     F. Beech, S. Mariglia, A. Santoro, and R.S. Roth, Phys. Rev. B 35:8778 (1987)
     ; J.J. Capponi, C. Chaillout, A.W. Hewat, P. Lejoy, M. Marezio, N. Nguyen, B.
     Raveau, J.L. Soubeyroux, J.L. Tholence, and R. Tournier, Europhys. Lett.
     3:1301 (1987).
2.   J.D. Jorgensen, M.A. Beno, D.G. Hinks, L. Soderholm, K.J. Volin, R.L.
     Hitterman, J.D. Grace, I.K. Schuller, C.U. Segre, K. Zhang, and S. Kleefisch,
     Phys. Rev. B 36:3608 (1987) ; J.D. Jorgensen, B.W. Veal, W.K. Kwok, G.W.
     Crabtree, A. Umezawa, L.J. Nowicki, and A.P. Paulikas, Phys. Rev. B 36:5731
     (1987).

3. The crystallographic site nomenclature used throughout this paper is that defined in Ref. 2.
4. P.K. Gallagher, H.M. O'Bryan, S.A. Sunshine, and D.W. Murphy, Mat. Res. Bull. 22:995 (1987) ; I.K. Schuller, D.G. Hinks, M.A. Beno, D.W. Cappone II, L. Soderholm, J.P. Locquet, Y. Bruynseraede, C.U. Segre, and K. Zhang, Solid St. Commun. 63:385 (1987).
5. J.M. Bell, Phys. Rev. B 37:541 (1988) ; L.T. Wille, A. Berera, and D. de Fontaine, Phys. Rev. Lett. 60:1065 (1988).
6. G. Van Tendeloo, H.W. Zandbergen, and S. Amelinckx, Solid St. Commun. 63:603 (1987) ; M.A. Alario-Franco, J.J. Capponi, C. Chaillout, J. Chenavas, and M. Marezio, in "High-Temperature Superconductors", M.B. Brodsky, ed., Material Research Society, Pitsburg, p. 41 (1988).
7. S. Amelinckx, G. Van Tendeloo, and J. Van Landuyt, in these Proceedings.
8. R.J. Cava, B. Batlogg, C.H. Chen, E.A. Rietman, S.M. Zahurak, and D. Weber, Phys. Rev. B 36:5719 (1987) ; W.E. Farneth, R.K. Bordia, E.M. McCarron III, M.K. Crawford, and R.B. Flippen, Solid St. Commun. 66:953 (1988).
9. J.M. Tranquada, D.E. Cox, W. Kunnmann, H. Moudden, G. Shirane, M. Suenaga, P. Zolliker, D. Vaknin, S.K. Sinha, M.S. Alvarez, A.J. Jacobson, and D.C. Johnston, Phys. Rev. Lett. 60:156 (1988) ; D.C. Johnston, S.K. Sinha, A.J. Jacobson, and J. Newsam, Physica C 153-155:572 (1988).
10. W.K. Kwok, G.W. Crabtree, A. Umezawa, B.W. Veal, J.D. Jorgensen, S.K. Malik, L.J. Nowicki, A.P. Paulikas, and L. Nunez, Phys. Rev. B 37:106 (1988) ; J.D. Jorgensen, H. Shaked, D.G. Hinks, B. Dabrowski, B.W. Veal, A.P. Paulikas, L.J. Nowicki, G.W. Crabtree, W.K. Kwok, L.H. Nunez, and H. Claus, Physica C 153-155:578 (1988).
11. J. Karpinski, E. Kaldis, E. Jilek, S. Rusiecki, and B. Bucher, Nature 336:660 (1988).
12. R. Haydock, V. Heine, and M.J. Kelly, J. Phys. C 5:2845 (1972).
13. L.F. Mattheiss and D.R. Hamann, Solid St. Commun. 63:395 (1987) ; S. Massidda, J. Yu, A.J. Freeman, and D.D. Koelling, Phys. Lett. A 122:192 (1987) ; H. Krakauer, W.E. Pickett, and R.E. Cohen, J. Supercond. 1:111 (1988).
14. W.M. Temmerman, Z. Szotek, P.J. Durham, G.M. Stocks, and P.A. Sterne, J. Phys. F 17:L319 (1987).
15. F. Herman, R.V. Kasowski, and W.Y. Hsu, Phys. Rev. B 36:6904 (1987) ; F. Herman, R.V. Kasowski, and W.Y. Hsu, in "Novel Superconductivity", S.A. Wolf and V.Z. Kresin, ed., Plenum, New York, p. 521 (1987).
16. B.A. Richert and R.E. Allen, in "High-Temperature Superconductors", M.B. Brodsky, ed., Material Research Society, Pitsburg, p. 463 (1988).
17. J. Zaanen, A.T. Paxton, O. Jepsen, and O.K. Andersen, Phys. Rev. Lett. 60:2685 (1988).
18. B.A. Richert and R.E. Allen, Japan. J. Appl. Phys. Suppl. 26-3:989 (1987) ; Phys. Rev. B 37:7496 (1988) ; Phys. Rev. B 37:7869 (1988).
19. W.A. Harrison, "Electronic Structure and the Properties of Solids", W.H. Freeman and Co., San Francisco (1980).
20. B.A. Richert and R.E. Allen, in "High-T_c Superconducting Films, Devices and Characterizations", G. Margaritondo, M. Onellion, and R. Joynt, ed., AIP Conference Series, American Institute of Physics, New York (in press, 1989).
21. P.O. Löwdin, J. Mol. Spectrosc. 14:112 (1964).
22. L.F. Mattheiss, Phys. Rev. B 6:4718 (1972).

23. For a detailed discussion of this band, see W.E. Pickett, Rev. Mod. Phys. (in press, 1989).
24. G. Allan, in "The Recursion Method and its Applications", D.G. Pettifor and D.L. Weaire, ed., Springer-Verlag, Berlin, p. 61 (1985).
25. Ph. Lambin and J.-P. Gaspard, J. Phys. F 10:651 (1980).
26. R.V. Kasowski, Superlatt. and Microstr. 3:383 (1987).
27. L.T. Wille and D. de Fontaine, Phys. Rev. B 37:2227 (1988) ; J.M. Sanchez and J.L. Morán-López, in these Proceedings.
28. M.A. Alario-Franco, C. Chaillout, J.J. Capponi, and J. Chenevas, Mater. Res. Bull. 22:1685 (1987).
29. J. Karpinski, E. Kaldis, S. Rusiecki, E. Jilek, P. Fisher, P. Bordet, C. Chaillout, J. Chenavas, J.L. Hodeau, and M. Marezio, in "Proceedings of the European MRS Meeting, Strasbourg, 8 - 11 Nov. 1988" (in press, 1989).
30. B. Domenges, M. Hervieu, C. Michel, and B. Raveau, Europhys. Lett. 4:211 (1987) ; H.W. Zandbergen, R. Gronski, K. Wang, and G. Thomas, Nature 351:596 (1988).
31. N.G. Stoffel, J.M. Tarascon, Y. Chang, M. Onellion, D.W. Niles, and G. Margaritondo, Phys. Rev. B 36:3986 (1987).
32. T. Fujiwara and Y. Hatsugai, Japan. J. Appl. Phys. 26:L716 (1987).
33. D.A. Papaconstantopoulos, W.E. Pickett, and M.J. De Weert, Phys. Rev. Lett. 61:211 (1988).
34. A.W. Hewat, J.J. Capponi, C. Chaillout, M. Marezio, and E.A. Hewat, Solid St. Commun. 64:301 (1987).
35. Referring to the densities of states of Figs. 2-4, which do not show a true energy gap around E_F+2 eV, the upper limit of the integrals leading to the hole numbers is not well-defined. However, the uncertainty of about 0.3 eV on the Cu(3d)-O(2p) band edge only yields a 0.01 uncertainty in the hole numbers given in Table 2.
36. Oxygen 2p hole counts in $YBa_2Cu_3O_x$, estimated from x-ray absorption spectroscopy, exhibit a similar monotonous x-dependent behavior : see e.g. P. Kuiper, G. Kruizinga, J. Ghijsen, M. Grioni, P.J.W. Weijs, F.M.F. de Groot, G.A. Sawatzky, H. Verweij, L.F. Feiner, and H. Petersen, Phys. Rev. B 38:6483 (1988).
37. J. van den Berg, C.J. van der Beek, P.H. Kes, G.J. Nieuwenhuys, J.A. Mydosh, H.W. Zandbergen, F.P.F. van Berkel, R. Steens, and D.J.W. Ijdo, Europhys. Lett. 4:737 (1987).
38. Y. Kitaoka, S. Hiramatsu, K. Ishida, T. Koharata, and K. Asayama, J. Phys. Soc. Japan 56:3024 (1987).
39. D. de Fontaine, M. Mann, and G. Ceder, in these Proceedings.
40. M. Hervieu, B. Domenges, C. Michel, and B. Raveau, Europhys. Lett. 4:205 (1987).
41. H.W. Zandbergen and G. Thomas, Phys. Status Solidi (a) 107:825 (1988).
42. G. Van Tendeloo, H.W. Zandbergen, and S. Amelinckx, Solid St. Commun. 63:389 (1987) ; J.C. Barry, J. Electr. Microsc. Techn. 8:325 (1988).
43. J.L. Hodeau, C. Chaillout, J.J. Capponi, and M. Marezio, Solid St. Commun. 64:1349 (1987).

ELECTRONIC STRUCTURE OF THE CUPRATE AND BISMUTHATE

SUPERCONDUCTORS

L. F. Mattheiss*

Condensed State Physics Research Department
AT&T Bell Laboratories
Murray Hill, N.J. 07974

The cuprate and bismuthate high-T_c superconductors share novel electronic properties which are significantly different from those exhibited by less exotic oxides. Most notable is the fact that E_F falls within the O(2p) band manifold, thereby creating a rare situation where O-derived electrons participate in the conduction process. Since the itinerant carriers have predominant O(2p) orbital character, it is not surprising that the electronic and superconducting properties of these high-T_c materials are especially sensitive to oxygen stoichiometry and disorder.

While the focus of this conference is concerned with the effects of oxygen disorder on the superconducting properties of the new oxide high-T_c superconductors, it is appropriate to consider the electronic properties of the idealized defect-free materials, since this is the practical starting point for understanding the effect of material imperfections on the electronic properties. In this paper, I plan to review the electronic band properties of the cuprate and bismuthate superconductors. In particular, I shall try to show how band theory can provide important insight into the basic chemistry of these materials, thereby suggesting ways in which their superconducting properties may be inproved.

I begin by comparing the electronic properties of the new high-T_c oxide superconductors with those exhibited by their more conventional low-T_c counterparts. The key differences in the electronic properties for these two classes of materials are shown schematically in Fig. 1. The high-T_c superconductors are characterized by nearly-degenerate Cu(3d)-O(2p) or Bi(6s)-O(2p) bands in which the Fermi level falls within the uppermost σ-

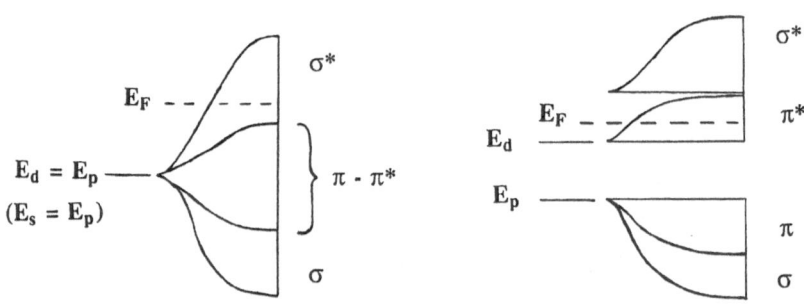

Fig. 1. Comparison of schematic bands for the high- and low-T_c oxide superconductors.

antibonding (or σ^*) subband, leaving it nearly half filled. The orbitals which form this subband in the cuprates are the Cu $d(x^2\text{-}y^2)$ and σ-antibonding O $p(x,y)$ states.[1] The Cu(3d) states are replaced by the Bi(6s) orbitals in the bismuthates.[2]

In the low-T_c materials, the metal d states lie above the O(2p) manifold, and E_F now occurs within the π^* or antibonding bands. The states at E_F have predominant d weight, in contrast to the high-T_c materials where there is comparable p-d (or s-p) admixture. Consequently, O stoichiometry and disorder are expected to be less important in the low-T_c versus the high-T_c materials. With E_F within the triply-degenerate π^* manifold, the density of states as well as the carrier density of the low-T_c compounds are generally higher than the corresponding values in the high-T_c materials.

An essential ingredient for determining the electronic structure of any material is a knowledge of its crystal structure. The primitive unit cells for all but the latest compounds are shown in Fig. 2. Some of the materials shown are the insulating parent compounds (BaBiO$_3$, La$_2$CuO$_4$) which must be doped appropriately before they superconduct. Others, such as 1:2:3 and the Tl and Bi cuprates, are "self doped" and superconduct in their stoichiometic composition.

In contrast to barium bismuthate, which forms with a distorted version of the cubic perovskite structure, the cuprates crystallize with a rich variety of layered perovskite-type structures. A characteristic feature of all these cuprates is the presence of one or more CuO$_2$ planes which are separated by various semimetallic or insulating layers. These CuO$_2$ layers represent the electronic "heart" of these cuprate superconductors. The primary role of the semimetallic and insulating layers is to adjust the band filling within the individual or

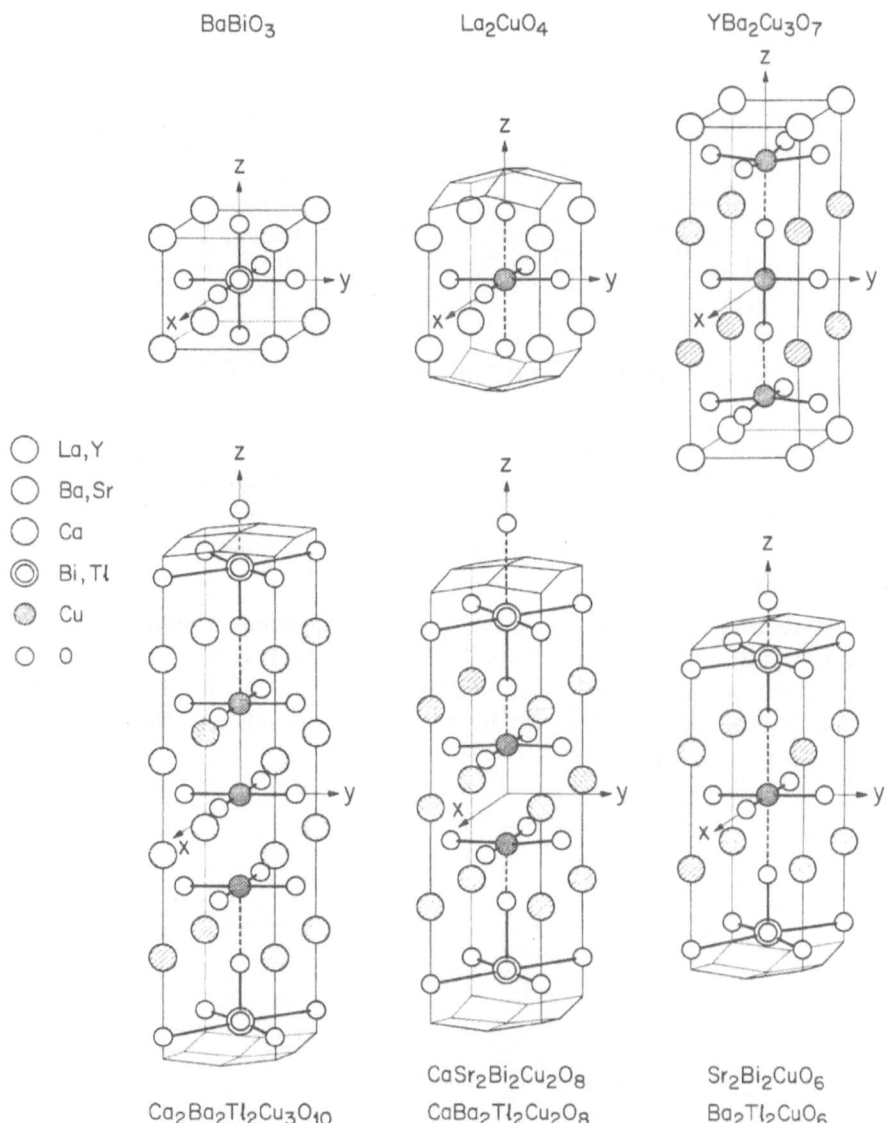

$BaBiO_3$ La_2CuO_4 $YBa_2Cu_3O_7$

La,Y
Ba,Sr
Ca
Bi,Tl
Cu
O

$Ca_2Ba_2Tl_2Cu_3O_{10}$ $CaSr_2Bi_2Cu_2O_8$ $Sr_2Bi_2CuO_6$
$CaBa_2Tl_2Cu_2O_8$ $Ba_2Tl_2CuO_6$

Fig. 2. Crystal structures of the bismuthate and cuprate superconductors.

multiple CuO_2 planes.

The insulating parent compound of the 1:2:3:7 phase is obtained by removing the central O's which form the 1-D chains along the y axis in Fig. 2. The corresponding parent of the Tl and Bi cuprates is more subtle. In these materials, one can insert additional CuO_2 and Ca layers, leading to increased

CaCuO₂   BaBiO₃

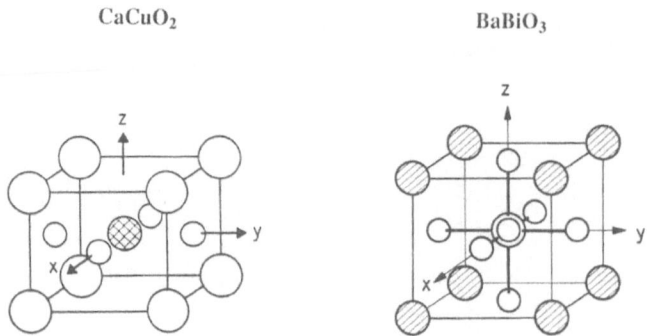

Fig. 3. Primitive unit cells for CaCuO$_2$ and BaBiO$_3$.

T_c's. The corresponding parent compound, CaCuO$_2$, is obtained when an infinite number of such layers is inserted.[3] The simplicity of its simple-tetragonal structure, which is shown in Fig. 3, makes it an useful prototype for comparing the electronic properties of the cuprate and bismuthate superconductors.

Local-density-approximation (LDA) electronic band-structure results for calcium cuprate[4] and barium bismuthate[5] are compared in Fig. 4 for wave vectors along comparable symmetry lines in the basal plane of the respective Brillouin zones. In CaCuO$_2$, we have a valence band consisting of nearly-

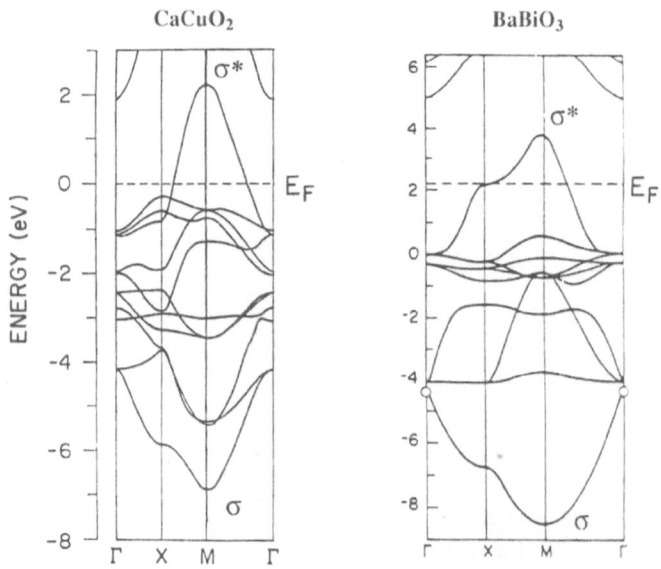

Fig. 4. LAPW energy-band results for CaCuO$_2$ and BaBiO$_3$.

degenerate Cu(3d) and O(2p) states. The corresponding valence-band complex in $BaBiO_3$ contains a single Bi(6s) state along with nine O(2p) bands. In both cases, the Ca- and Ba-derived bands lie well above E_F so they are minimally admixed with the valence-band manifold. Thus, Ca and Ba are electronically inactive (donor) constituents.

The Fermi level cuts the uppermost σ-antibonding subband in both materials, leaving it exactly half filled. In each case, the orbitals that form the half filled subbands are those which form σ bonds with neighboring O's. In the cuprates, the key Cu(3d) orbital is the one with $d(x^2-y^2)$ symmetry. This state forms σ bonds with the four neighboring O(2p) orbitals that point along the Cu-O bond direction. In barium bismuthate, the $d(x^2-y^2)$ orbital is replaced by the more extended Bi(6s) state, which now forms σ bonds with its six O neighbors. Using tight-binding models, one can estimate the relative strengths of these [(pdσ)~1.6 eV and (spσ)~2.2 eV] interactions.[1,2]

This half-filled band condition in the cuprate and bismuthate parent compounds produces an electronic instability.[1,2] This instability is removed differently in the two classes of materials. In the bismuthates, it causes a Peierls distortion in which the O's surrounding one Bi site are displaced toward its neighbor. This so-called "breathing-mode" displacement of the oxygens doubles the size of the primitive unit cell and halves the size of the corresponding Brillouin zone. The effect of the distortion on the conduction-band states near E_F is shown in Fig. 5.[2,5] The cubic bands, which are degenerate within the folding planes (W-L) of the Brillouin zone, are now split

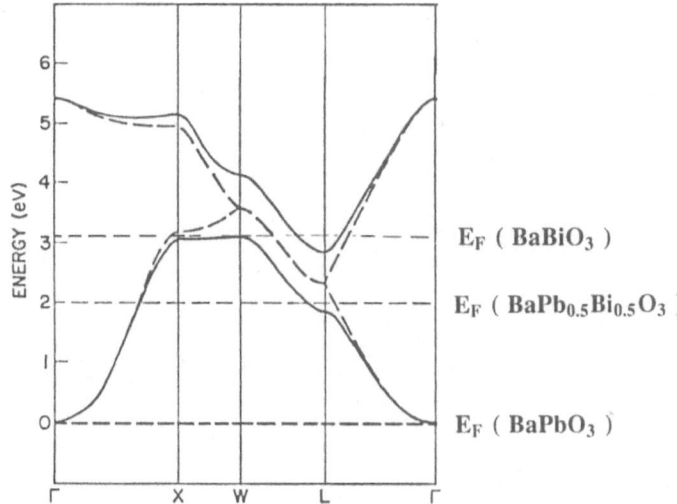

Fig. 5. Model tight-binding conduction bands for cubic (dashed) and distorted (solid) $BaPb_{1-x}Bi_xO_3$ alloys.

by the breathing-mode distortion. This lowers the average band energy of the states near E_F and stabilizes the distortion. The origin of these splittings can be traced[2] to the modulation of the (spσ) interaction as a given oxygen moves closer to one Bi than the other. The role of doping is to inhibit the frozen-in commensurate distortion, thereby producing a metal in which the conduction electrons are strongly coupled to the dynamic O breathing-type phonons.

In the case of the cuprates, the story is different. Here, the electronic instability produces an antiferromagnetic insulator[6], with alternating spins on neighboring Cu sites within the CuO_2 planes. Attempts to explain this AF behavior in terms of itinerant antiferromagnetism have been unsuccessful.[7] This has led to the proposal[8,9] that band theory is inadequate for describing the electronic properties of the cuprates. In particular, it has been suggested[8,9] that e-e correlations are an essential ingredient for understanding the basic electronic structure of these materials.

The effect of e-e correlations on the one-electron LDA band picture is shown schematically in Fig. 6. The hybridized Cu(3d)-O(2p) valence-band density of states is shown to the left, with the half-filled σ* subband. When e-e correlations are introduced via a Hubbard-type model[9], the Cu(3d) electrons localize, forming a pair of Hubbard subbands which are separated by the Coulomb energy U_d. The lower $3d^9$ Hubbard band is occupied and separated by an energy ε from the centroid of the O(2p) manifold. Thus, according to this model, one can understand why the parent compounds are semiconductors rather than metals. Furthermore, when extra holes are introduced into the

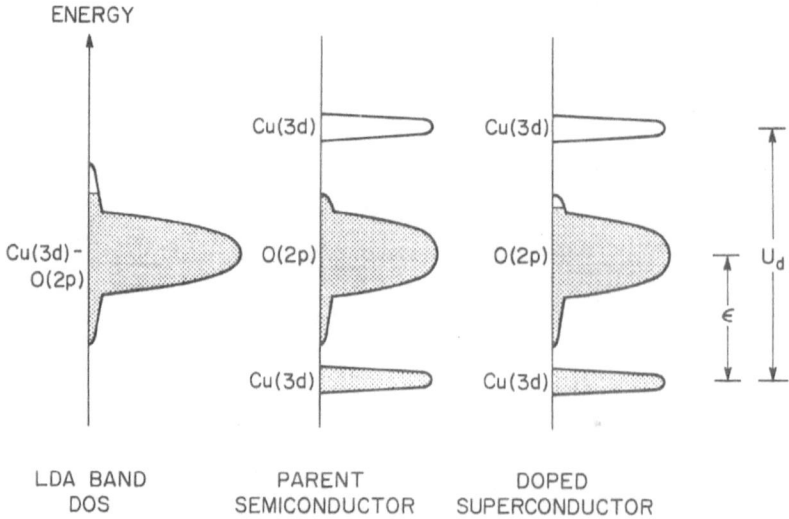

Fig. 6. Comparison of LDA band density-of-states (DOS) results with those derived from a Hubbard-type treatment of electron-electron correlations.

U_d, producing filled and empty Hubbard subbands with intermediate O(2p) bands. Again, the stoichiometric compound is an insulator. However, any additional holes that are introduced into the system will depopulate the uppermost $\sigma(x,y)$ bands (i.e. O(2p) states that form σ bonds with the Cu $d(x^2-y^2)$ orbitals). The $\pi(x,y)$ and (dashed) $\pi(z)$ bands begin to enter only at larger hole concentrations.

We consider next the bismuthates. The original procedure[10] for adjusting the band filling in $BaBiO_3$' involved Pb doping at the active Bi site. The calculated bands[5] for Pb-doped barium bismuthate alloy system are compared in Fig. 8. Included are the end members, along with one material with intermediate composition which is calculated in the "virtual-crystal" approximation. As shown, Pb doping can adjust the filling of the σ-antibonding subband, but it also introduces a noticeable chemical shift in the position of the 6s (open circles) states. In all cases, the Ba bands are well above E_F.

Based on this simple band picture, one might expect the metallic properties of these materials to extend over most of the intermediate range of alloys, at least until the band filling was close to that of the Peierls-distorted-insulator, barium bismuthate. In fact, the observed phase diagram[11] shows that these alloys are metallic and superconducting (with a maximum $T_c \approx 12K$) for low Bi concentrations, but then become semiconducting for $x > 0.35$.

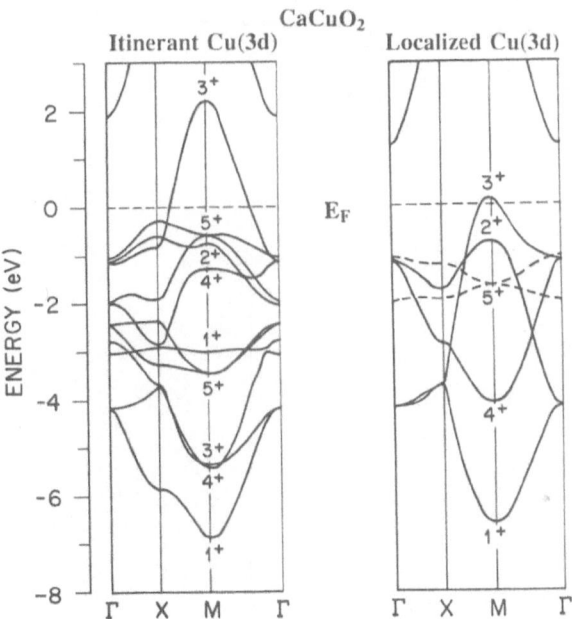

Fig. 7. LAPW bands for $CaCuO_2$ with itinerant and localized Cu(3d) electrons.

Though the energy of the 3d level is not determined by this procedure, tight-binding estimates[4] show that is located in the energy range of -2 to -3 eV.

As in Fig. 6, we can now imagine turning on the on-site Coulomb energy system, they depopulate the previously filled O(2p) bands, producing metallic behavior as well as high-temperature superconductivity.

This schematic picture can be made more concrete by means of model linear augmented-plane-wave (LAPW) calulations on a prototype cuprate in which the Cu(3d) electrons have been artificially localized.[4] The basic idea is very simple. We essentially treat the Cu(3d) electrons as core rather than valence states within the framework of the LAPW method. More specifically, we use a pseudopotential trick to orthogonalize the LAPW basis to the Cu(3d) states.

The effect of this procedure on the previously discussed calcium cuprate band structure is shown in Fig. 7. To the left are the standard LAPW valence bands with itinerant Cu(3d) electrons. These are identical to the results shown in Fig. 4. The bands to the right are those calculated in the limit where the Cu(3d) electrons are localized. Notice now that the 3d bands are missing, leaving only the six O(2p) bands. The labels allow one to see how the various states are shifted in the transition from the itinerant to the localized 3d limit.

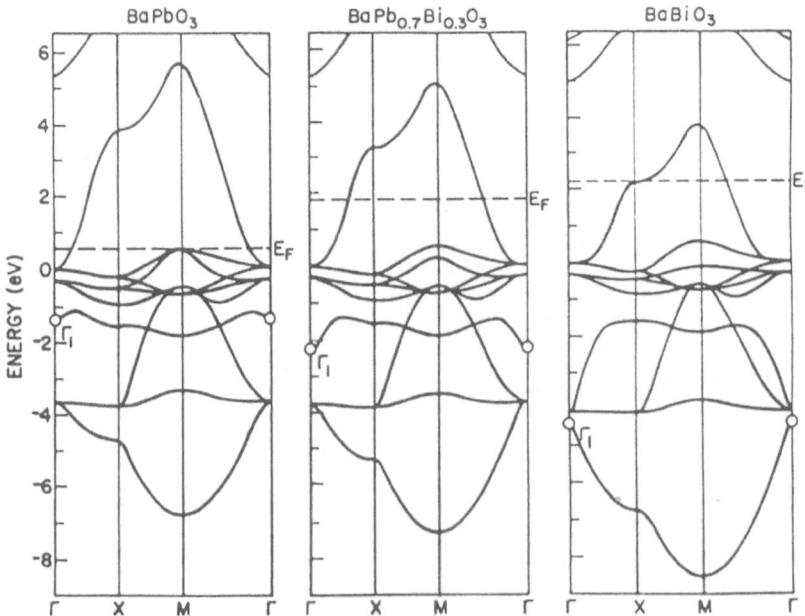

Fig. 8. LAPW bands for $BaPb_{1-x}Bi_xO_3$ alloys.

Simple theoretical ideas based on a phonon-mediated mechanism suggest that T_c would continue to increase if the metallic phase could be extended closer to half filling. This can be understood in terms of the simple band picture shown in Fig. 5. There, we compare the undistorted conduction bands (solid) with those (dashed) derived from a frozen-in breathing-mode distortion of oxygens. A measure of the electron-phonon coupling strength is represented by the shift of the dashed-to-solid energy-band states. These shifts increase monatonically with band filling, and reach a maximum at half filling, barium bismuthate. Of course, here the coupling strength is such that it induces a static distortion.

The objective is to achieve a marginally stable situation where the coupling is large and the static distortion is avoided. According to model calculations by Weber[12], one can explain the premature semiconducting behavior in $BaPb_{1-x}Bi_xO_3$ in terms of Pb-Bi ordering and incommensurate distortions at the electronically active Bi sites. This suggested[13] the advantages of an alternate doping strategy in which the band filling was adjusted by means of alkali doping on the inactive Ba sites. This has led to the discovery[13,14] of a 40% K-doped barium bismuthate material that remains cubic and metallic within the previous semiconducting regime and exhibits a superconducting transition near 30K.

REFERENCES

* This work has been done in collaboration with D. R. Hamann

1. L. F. Mattheiss, Phys. Rev. Lett. 58, 1028 (1987).
2. L. F. Mattheiss and D. R. Hamann, Phys. Rev. B 28, 4227 (1983).
3. T. Siegrist et al., Nature (London) 334, 231 (1988).
4. L. F. Mattheiss and D. R. Hamann, Phys. Rev. B (to be published).
5. L. F. Mattheiss, Jpn. J. Appl. Phys. 24, Suppl. 2, 6 (1985).
6. D. Vaknin et al., Phys. Rev. B (to be published).
7. T. C. Leung, X. W. Wang, and B. N. Harmon, Phys. Rev. B 37, 384 (1988); P. A. Sterne and C. S. Wang, Phys. Rev. B 37, 7472 (1988).
8. P. W. Anderson, Science 235, 1196 (1987).
9. V. J. Emery, Phys. Rev. Lett. 58, 2794 (1987).
10. A. W. Sleight, J. L. Gillson, and P. E. Bierstedt, Solid State Commun. 17, 27 (1975).
11. S. Uchida, K. Kitazawa, and S. Tanaka, Phase Transitions B 8, 95 (1987).
12. W. Weber, Jpn. J. Appl. Phys. 26, Suppl. 3, 981 (1987).
13. L. F. Mattheiss, E. M. Gyorgy, and D. W. Johnson, Jr., Phys. Rev. B 37, 3745 (1988).
14. R. J. Cava et al., Nature (London) 332, 814 (1988).

FLUX PINNING AND THERMALLY ACTIVATED FLUX FLOW BEHAVIOR IN SINGLE CRYSTALLINE YBa$_2$Cu$_3$O$_7$ AND Bi$_2$Sr$_2$CaCu$_2$O$_8$

J. van den Berg, P.H. Kes, P. Koorevaar, G.J. Nieuwenhuys
J.A. Mydosh, M.J.V. Menken* and A.A. Menovsky*

Kamerlingh Onnes Laboratory, State University Leiden
P.O. box 9506, 2300 RA Leiden, the Netherlands and
*Physics Laboratory of the University of Amsterdam

ABSTRACT

In this paper we will focus on some magnetic properties of single crystalline YBa$_2$Cu$_3$O$_7$ and Bi$_2$Sr$_2$CaCu$_2$O$_8$. Using Bean's critical state model, critical–current densities and flux pin forces are determined from magnetization versus magnetic field loops. In YBa$_2$Cu$_3$O$_7$ we will describe the flux pin behavior at lower temperatures in terms of flux–line pinning by twin planes. In Bi$_2$Sr$_2$CaCu$_2$O$_8$ the flux pin forces are one order of magnitude smaller than in YBa$_2$Cu$_3$O$_7$, due to the absence of twin planes. Therefore, thermally activated depinning is much more pronounced in Bi$_2$Sr$_2$CaCu$_2$O$_8$. After a theoretical introduction we will apply the thermally activated flux flow model (TAFF) on our magnetic measurements of Bi$_2$Sr$_2$CaCu$_2$O$_8$. The time dependent response of the magnetization on a small step in the magnetic field can be explained by TAFF. The predicted relation between the typical relaxation time after a field step and the critical current density in presence of thermally activated depinning, J_c, is confirmed. Furthermore, the reversible magnetization curve can be measured at higher temperatures due to thermally activated depinning. Estimations for H_{c1} are thus obtained.

1. INTRODUCTION

For technical applications of superconductors, several properties are of vital importance. With the discovery of the high temperature oxide superconductors with transition temperatures above liquid nitrogen temperature (77 K), an enormous leap forward has been made with respect to the problem of cooling. However, a high transition temperature is not enough, also the upper

critical magnetic field H_{c2} and the critical current density J_c of the superconducting material have to be high. The experimental data now available yield
a very high anisotropic upper critical field, due to the very small anisotropic
Ginzburg–Landau coherence lengths ξ_{ab} and ξ_c. From pulsed field magnetization measurements we determine for a collection of $YBa_2Cu_3O_7$ single crystals,
$H_{c2}(0) \approx 60$ T for magnetic fields parallel to the c–axis (see below).

The critical current density of a superconductor is an extrinsic property.
In a magnetic field larger than the lower critical field H_{c1}, flux lines are present
in the sample, which at higher fields form an hexagonal flux–line lattice. Move-
ment of the flux lines under influence of a driving force, e.g. a current through
the sample or a gradient in the magnetic field, is a dissipative process and as
a consequence, losses will occur. Fortunately, the flux lines are pinned by defects in the crystal, like voids, quasi–dislocation loops and stacking faults. The
magnetic flux lines will be pinned, until the driving force exceeds the pinning
force. This will be experimentally observable by a voltage over the sample or
a time dependence in the magnetic response. In this article the elementary
pinning force of a planar defect (grain boundary, twin plane) on a flux line
will be calculated and compared with the experimentally determined values for
$YBa_2Cu_3O_7$.

In $Bi_2Sr_2CaCu_2O_8$ single crystals, a strong dependence of the critical current density on temperature and magnetic field is observed, as well as a striking
frequency dependent decrease of the diamagnetic onset temperature, determined by ac susceptibility in a static magnetic field [1]. These phenomena are
attributed to thermally activated processes, as was first suggested by Malozemoff et al. [2]. Kes et al. [3] described these processes by solving a continuity
equation for flux lines. They show that for the limit of small driving forces the
continuity equation reduces to a simple diffusion equation, which is easily solved
for several boundary conditions. To further investigate the TAFF model, we
have checked the predicted relation between the critical current density and the
typical relaxation time for the magnetization after a small step in the magnetic
field.

Due to the thermally activated depinning of flux lines, it is possible to
determine the reversible magnetization curve at higher temperatures, which in
principle reveals H_{c1} and, from the slope of $M(H)$ at H_{c2}, the parameter κ.
Being limited by magnetic field strength and sensitivity, we can only obtain an
estimation of H_{c1}.

2. EXPERIMENTAL TECHNIQUES

Magnetization measurements were performed using a Foner–type vibrat-
ing sample magnetometer (VSM), suitable for the temperature range from 1.5
to 300 K and for magnetic fields up to 5 T. The advantage of using a VSM
over a SQUID magnetometer is that after a change in the magnetic field the
magnetization can be determined much faster with a VSM, but the sensitivity

is less. Magnetization up to 35 T was determined with a pulsed field set up, at temperatures of liquid He, H_2 and N_2.

3. THEORY

3.1. Flux Pinning by Twin Planes

To calculate the elementary pinning force of a twin plane, we use the theory originally developed for pinning on grain boundaries by Thuneberg [4,5]. In this theory pinning due to local variation of the mean free path l_{tr} is calculated. Thuneberg [4] has shown that in most cases this so called $\delta\kappa$ effect outweighs the pinning due to local variation in T_c, the so called δT_c effect, by at least a factor of 10. If the twin planes in $YBa_2Cu_3O_7$ contribute to extra electron scattering, the theory should be applicable for this material as well.

Furthermore, the method of Pruymboom [6] will be used to extend the computations to lower temperatures. Eventually the elementary pinning force will be determined for all temperatures and magnetic fields.

We start with the Ginzburg–Landau functional in non–reduced units, in the de Gennes notation [7]:

$$\mathcal{F} = \int d\vec{r}\left\{A|\Delta|^2 + \frac{B}{2}|\Delta|^4 + C|\partial\Delta|^2 + \frac{h^2}{2\mu_0}\right\} \tag{3.1}$$

$$\vec{\partial} \equiv i\vec{\nabla} - (\frac{2e}{\hbar})\vec{A} \tag{3.2}$$

Δ is the gap function, proportional to the order parameter. The motivation to work with non–reduced quantities becomes clear when we write the coefficients A, B and C:

$$A = N(0)(\frac{T}{T_c} - 1) \tag{3.3}$$

$$B = 0.098\frac{N(0)}{(kT_c)^2} \tag{3.4}$$

$$C = 0.55\xi_0^2 N(0)\chi(\alpha), \ \alpha = 0.882\frac{\xi_0}{l_{tr}} \ . \tag{3.5}$$

Here $\chi(\alpha)$ is the Gorkov impurity function, and α is the impurity parameter. l_{tr} is the mean free path of the electrons. We have now splitted the δT_c and the $\delta \kappa$ contributions: the coefficients A and B are dependent on T_c, and C on l_{tr} only. The extra electron scattering influences the mean free path, and therefore the coefficient C. Taking into account only the $\delta \kappa$ effect, the pin potential is given by the variational of the free energy with respect to perturbations of the mean free path by the defect:

$$\delta_l \Omega(\vec{r}) = c(T) \int d\vec{r'} |\partial \Delta|^2 \delta \left(\frac{1}{l_{tr}(\vec{r} - \vec{r'})} \right) . \tag{3.6}$$

For a planar defect parallel to the flux lines, we assume the variational of the mean free path to be exponentially dependent on the distance to the defect, located at x' [6]:

$$\delta(\frac{1}{l_{tr}(x)}) = \frac{p_{tr}}{2r_0} \exp \left(-\frac{|x' - x|}{r_0} \right) . \tag{3.7}$$

p_{tr} denotes the scattering probability of the defect, and r_0 is the non-locality of the order parameter $(r_0^{-1} = \xi^{-1} + l_{tr}^{-1})$.

For the limit of isolated vortices $(b < 0.2)$, the expression given by Clem [8] is taken for the gap function:

$$|\Delta|^2 = \frac{\hbar^2}{4mC} \psi_\infty^2 |f|^2 \ , \ f = \frac{\rho}{\sqrt{\rho^2 + \xi^2}} \ , \tag{3.8}$$

with ρ the distance to the center of the vortex, and ψ_∞ the value of the order parameter in absence of a magnetic field. For $b \to 1$, the mixed state, we use:

$$|\Delta|^2 = \frac{\hbar^2}{4mC} \psi_\infty^2 \frac{(1-b)}{\beta_A} \left[1 - 2e^{-\frac{\pi}{\sqrt{3}}} \left\{ \cos(x - \frac{y}{\sqrt{3}}) \frac{2\pi}{a_0} + \right. \right.$$
$$\left. \left. + \cos(\frac{4\pi y}{\sqrt{3} a_0}) + \cos(x + \frac{y}{\sqrt{3}}) \frac{2\pi}{a_0} \right\} \right] . \tag{3.9}$$

a_0 is the distance between flux lines given by $a_0 = 1.075 \sqrt{\frac{\phi_0}{B}}$, with ϕ_0 the elementary flux quantum, and $\beta_A = 1.16$ is the Abrikosov parameter. After performing the integration of Eq.3.6, the maximum pinning force is given by the maximum in the derivative of $\delta_l\Omega$ with respect to x:

$$f_l(b) = \tilde{f}_l(b)\left\{g(\alpha)\mu_0 H_c^2 \xi_0 p_{tr}(\frac{\xi}{r_0})^2\right\} \tag{3.10}$$

$$\tilde{f}_l(b) = \frac{(1-b)}{\left\{(\frac{a_0}{r_0})^2 + \frac{16\pi^2}{3}\right\}} \tag{3.11}$$

$$g(\alpha) = \frac{-0.882\frac{\partial\chi(\alpha)}{\partial\alpha}}{\chi(\alpha).} \tag{3.12}$$

This is the microscopic pinning force, per flux line, per unit length along the twin plane.

To calculate the macroscopic pinning force F_p from the microscopic pinning force we use a direct summation method:

$$F_p = \frac{f_l}{La_0}, \tag{3.13}$$

where L is the mean distance between twin planes. It should be noted here that the twin planes can exert a pinning force perpendicular to the twin plane only. Therefore, perpendicular twin planes are necessary to contribute to the total pinning force.

3.2. The Thermally Activated Flux Flow Model

The TAFF model, which describes the thermally activated motion of flux lines over potential barriers, is extensively described in ref. 3. We will here give a short introduction only. It is shown in ref. 3 that the continuity equation for flux bundles in a periodic potential reduces to a simple diffusion equation in the limit of small driving forces. For small variations in the magnetic field $B(r) = B_0 + \delta B(r)$ the induction in a cylinder is described by:

$$\frac{\partial B}{\partial t} = \frac{D_0}{r}\frac{\partial}{\partial r}\left(r\frac{\partial B}{\partial r}\right), \tag{3.14}$$

with the diffusion constant D_0 given by:

$$D_0 = \frac{2\nu_0 B_0 w^2 V_c}{\mu_0 kT}\exp\left(-\frac{U}{kT}\right). \tag{3.15}$$

Here V_c is the volume of an elastically independent flux bundle, taken to be the correlated volume from the theory of collective flux pinning [9]. Furthermore, U is the height of the pin potential, ν_0 is an attempt frequency and w is the average jump distance of V_c.

The diffusion equation can now be solved for different boundary conditions.

- $H = H_{c2} - \dot{H}t$. With this boundary condition, the magnetization can be calculated for a cylinder with radius a:

$$M(t) = \frac{\dot{H}a^2}{8D_0}\left\{1 + 32\sum_{n=1}^{\infty}\left[\frac{-\exp(D_0\alpha_n^2 t)}{\alpha_n^4}\right]\right\} \tag{3.16}$$

$\dot{H}$ is the sweep rate of the magnetic field, and the α_n's are the positive roots of $J_0(\alpha a)$, where J_0 is the zeroth order Bessel function. For the limit of small $\dot{H}$ and large t, we obtain for a high-κ superconductor: $M = \frac{\dot{H}a^2}{8D_0}$. Using the Bean model, where $J_c = \frac{3M}{a}$, a relation between J_c and D_0 is obtained: $J_c = \frac{3\dot{H}a}{8D_0}$. Note that $J_c \neq J_{c0}$, the critical current density without TAFF. In the TAFF model, the "Irreversibility Line" [2] is not an intrinsic property, but dependent on the sweep rate of the magnetic field, and the sensitivity of the experimental set up.

- $H_0 \rightarrow H_0 + \delta H$. Solving the diffusion equation with this boundary condition yields the time dependence of the magnetization, after a small step δH in the magnetic field:

$$\delta M(t) = 4\delta H \sum_{n=1}^{\infty} \frac{\exp\left(-\frac{t}{\tau_n}\right)}{\beta_n^2} \tag{3.17}$$

with $\tau_n = \frac{\tau_0}{\beta_n^2}$, where $\tau_0 = \frac{a^2}{D_0}$ is the typical relaxation time, and $\beta_n = \alpha_n a$.

Thus, the TAFF theory predicts a relation between the measured critical current density J_c and the typical relaxation time after a field step, τ_0, via D_0:

$$\tau_0 = \frac{8a}{3\dot{H}}J_c . \tag{3.18}$$

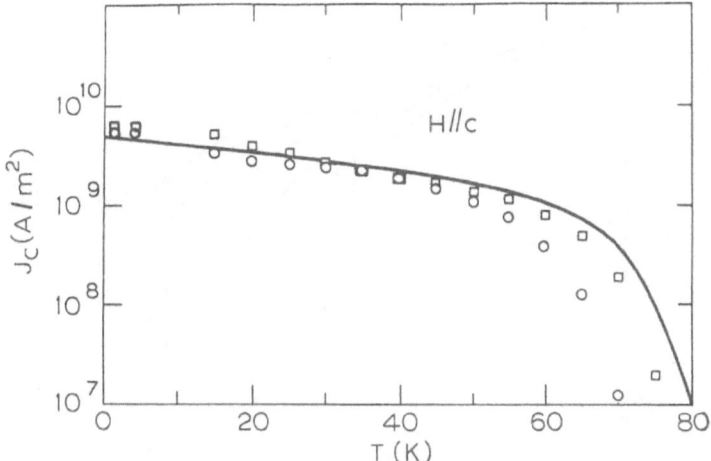

Fig 1. *Critical current density of a* $YBa_2Cu_3O_7$ *single crystal vs. temperature. Open squares and circles denote measurements at 2 and 4 T respectively, with the magnetic field directed parallel to the crystallographic c–axis. The solid line represents a fit to the TAFF model (see text).*

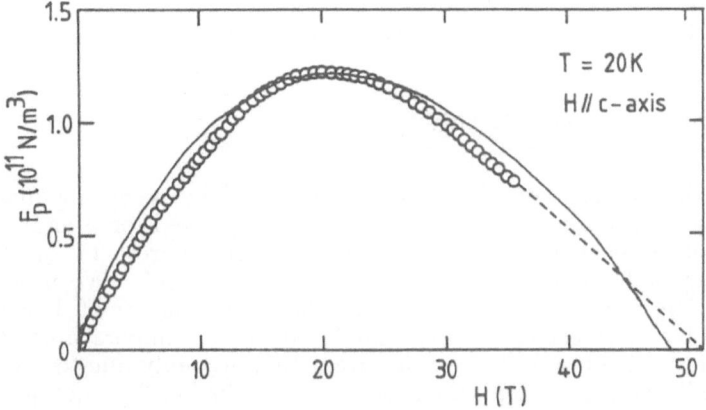

Fig 2. *Macroscopic pinning force density of a collection of c–axis aligned* $YBa_2Cu_3O_7$ *single crystals at 20 K. The dashed line represents a linear extrapolation, and the solid line is the scaled dome–shaped curve predicted by the TAFF model (see text).*

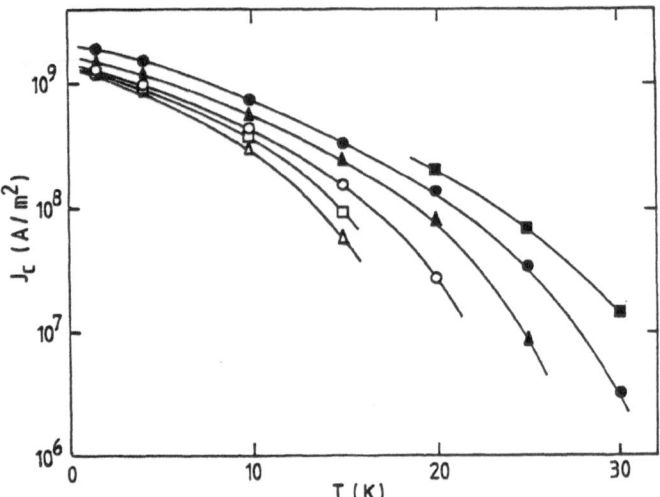

Fig 3. *Critical current density of a* $Bi_2Sr_2CaCu_2O_8$ *single crystal vs. temperature. Filled squares represent measurements at 0.2 T, filled circles at 0.5 T, filled triangles at 1 T, open circles at 2 T, open squares at 3 T and open triangles at 4 T. The solid lines are a guide to the eye only.*

4. EXPERIMENTAL RESULTS AND DISCUSSION

4.1. Measurements on $YBa_2Cu_3O_7$

Critical current densities were determined by recording magnetization loops at several temperatures. The results are plotted for 2 and 4 Tesla as a function of temperature in fig. 1 [10]. The solid line represents $J_c(T)$, predicted by pinning of twin–planes in the isolated vortex limit. We used $p_{tr} = 0.4$ and $L = 100$ nm. The curve is shifted down by a factor of 3, which can be explained by the uncertainty in p_{tr} and L. At higher temperatures, J_c decreases more rapidly with temperature than predicted, probably due to the occurrence of thermally activated depinning (see below). In fig. 2 a measurement on a c–axis aligned collection of single crystals is presented [11]. The macroscopic pinning force is calculated from the critical current density via $F_p = J_c \times B$. The measurement was performed in the pulsed field set up of the University of Amsterdam. The solid line is the dome–shaped curve, expected from Eq.3.13, scaled by a factor 2. Extrapolating the dome shape yields $H_{c2}(20K) = 49$ T, whereas a linear extrapolation yields $H_{c2}(20K) = 51$ T. Applying the WHH

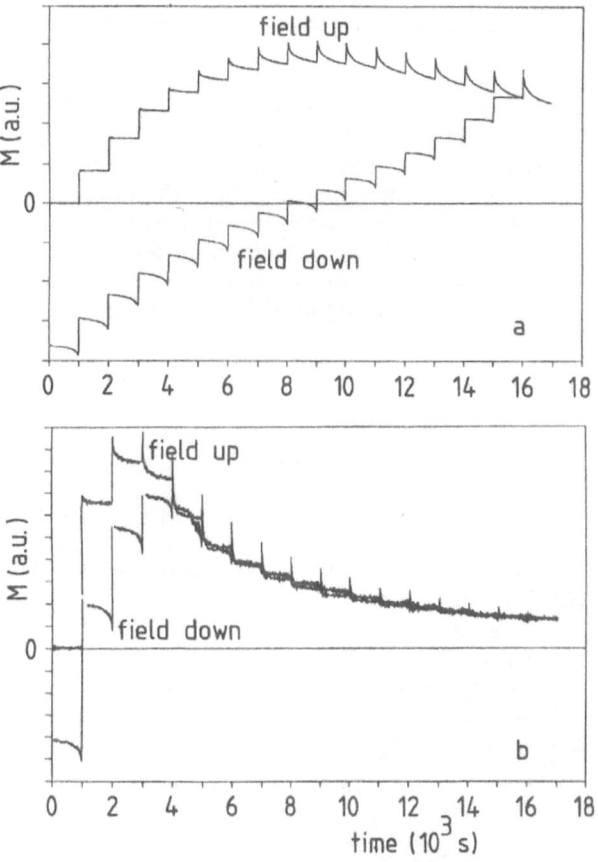

Fig 4. Magnetization vs. time at temperatures of 30 K (a) and 60 K (b) of a $Bi_2Sr_2CaCu_2O_8$ *single crystal. Each 1000 s the magnetic field is increased with 1 mT.*

[12] theory predicts $H_{c2}(0) \approx 60$ T, from which a Ginzburg–Landau coherence length of $\xi_{ab}(0) \approx 2.0$ nm, and a BCS coherence length in the clean limit of $\xi_0 \approx 2.7$ nm can be calculated. These values are very acceptable, and the twin planes seem to be a major pinning source, although a contribution to the total pinning force of point pins can not be excluded. The real proof for the pinning properties of twin planes is, however, in the decoration experiments by Dolan et al. [13].

4.2. <u>Measurements on $Bi_2Sr_2CaCu_2O_8$</u>

We will focus here on the magnetization measurements only. The ac susceptibility measurements are presented elsewhere [1], and reveal a diamagnetic onset temperature of 86 K and a sharp transition to the superconducting state. The calculated critical–current densities are given for several magnetic fields in fig. 3. The difference in temperature– and field dependence of J_c with respect to $YBa_2Cu_3O_7$ is remarkable. We attribute this to the absence of twin planes in $Bi_2Sr_2CaCu_2O_8$. Because of the smaller pinning forces in $Bi_2Sr_2CaCu_2O_8$, the effects of TAFF are much more pronounced. To study these effects, the response to small steps in the magnetic field was determined and compared with the prediction by the TAFF model.

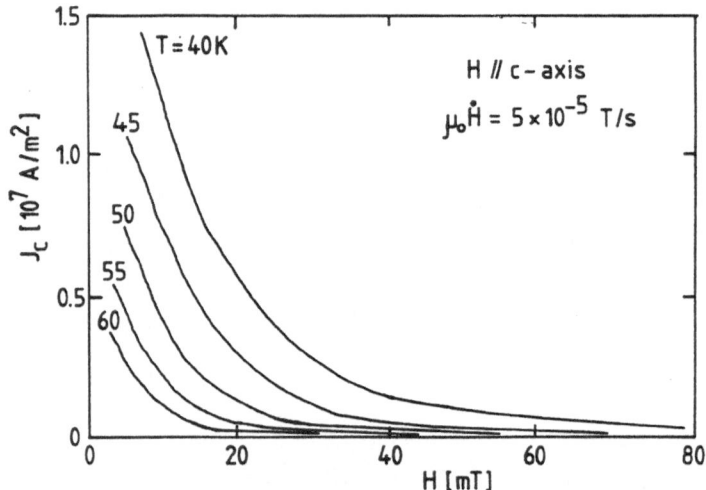

Fig. 5. Low–Field critical current density vs. magnetic field for several temperatures of a $Bi_2Sr_2CaCu_2O_8$ single crystal.

The experiment was performed as follows. After zero–field cooling the sample, the magnetic field was raised with steps of 1 mT, and the magnetization was recorded for 1000 s after each step. The results are given for temperatures of 30 (a) and 60 K (b) in fig. 4 and for fields up to 16 mT. At 30 K a 100 % shielding is observed for fields up to $3 \sim 4$ mT. For higher fields, above H_{c1}, flux start leaking into the sample, and a time dependent magnetization is observed. At 60 K, this time dependence occurs already at 1 mT, and above 4 mT, the motion of flux into the sample is so fast, that the magnetization becomes reversible. From this reversible magnetization curve, H_{c1} can be determined. Using a similar measurement at 70 K, we estimate $H_{c1}(70K)$ at 3 mT, which gives $H_{c1}(0) \approx 11mT$ [12]. This value is in accordance with the value reported by Lin et al. [14].

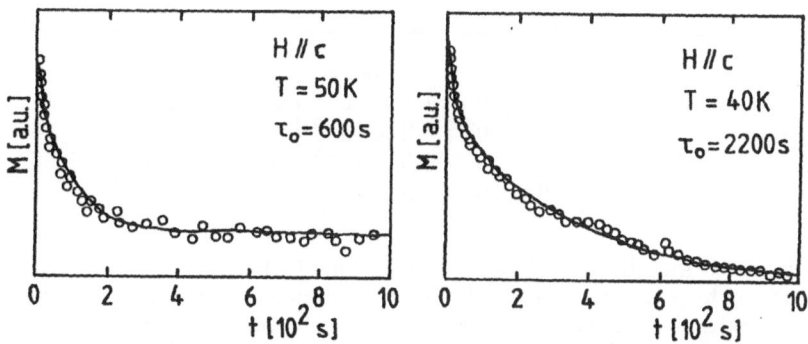

Fig. 6. Relaxations of the magnetization of a $Bi_2Sr_2CaCu_2O_8$ *single crystal, at temperatures of 40 and 50 K after a field step* $15 \rightarrow 16\ mT$. *The solid lines are fits to the TAFF model (see text).*

Unfortunately, our measurements are not accurate enough to fit the time dependent response to Eq.3.17 directly, but we can compare $M(t)$ with the behavior predicted by TAFF from high-temperature, low-field critical current measurements (see Eq.3.18). In fig. 5, J_c as a function of magnetic field is plotted for temperatures of 40–60 K and fields up to 80 mT. In Table 1, values for J_c, D_0, τ_0 and U are listed as a function of T for a magnetic field of 15 mT. In calculating U we used $V_c = 50a_0^3$, $\nu_0 = 10^{10}$ Hz and $w = \xi$ [1,3]. We analyze the specific measurements at 15 mT to stay within the limits of applicability of the TAFF model. When compared with the actual relaxation measurements, the calculated values for τ_0 are found to be a factor of 4 too small. Applying this factor 4 yields acceptable results for relaxations in the temperature range of 40–60 K. The proportionality between J_c and τ_0 implies that the TAFF model is self consistent. The factor 4 can easily be explained by the rough approximations made with respect to the sample geometry. In Table 1 this factor 4 is already included. In fig. 6 the recorded magnetization after a step in the magnetic field $H = 15 \rightarrow 16$ mT is plotted, as well as the relaxation predicted via the J_c measurements (Eq.3.17,3.18).

We find that $\frac{U}{kT}$ is only very slowly varying with T, which is very unusual. This is probably an artifact due to the assumptions made for V_c and ν_0.

5. CONCLUSIONS

In this paper we have presented some magnetic measurements on single crystalline $YBa_2Cu_3O_7$ and $Bi_2Sr_2CaCu_2O_8$. The low temperature measurements on $YBa_2Cu_3O_7$ can be explained by strong flux pinning of twin planes. At higher temperatures thermally activated depinning plays an important role. We have studied these effects in $Bi_2Sr_2CaCu_2O_8$, where the pinning forces are

Table 1

T [K]	$J_c[A/m^2]$	$D_0[m^2/s]$	$\tau_0[s]$	U/kT
40	8.1×10^6	1.8×10^{-9}	2113	22.3
45	4.6×10^6	3.2×10^{-9}	1200	21.8
50	2.3×10^6	6.5×10^{-9}	600	21.1
55	1.0×10^6	1.5×10^{-8}	261	20.3

smaller due to the absence of twin planes. The TAFF model describes correctly the relation between the time dependent magnetic response on a step in the magnetic field and the critical current measurements. To obtain reliable values for the pin potential U, however, more experimental as well as theoretical work to determine V_c and ν_0 is needed. At higher temperatures the reversible magnetization can be measured, as a consequence of thermally activated depinning. From this reversible magnetization we obtained $H_{c1}(0) = 11$ mT for H directed parallel to the crystallographic c–axis.

ACKNOWLEDGEMENTS

These investigations were financially supported by the Foundation for Fundamental Research on Matter (FOM) and were part of the Amsterdam Leiden Cooperation for Materials Research (ALMOS).

REFERENCES

[1] J. van den Berg, C.J. van der Beek, P.H. Kes, J.A. Mydosh, M.J.V. Menken and A.A. Menovsky, Supercond. Sci. Technol. 1:249 (1989).

[2] For an excellent overview see: Macroscopic Magnetic properties of High Temperature Superconductors, A.P. Malozemoff, to be published in: "Physical Properties of High Temperature Superconductors," D. Ginsberg, ed., World Scientific, Singapore, 1989.

[3] P.H. Kes, J. Aarts, J. van den Berg, C.J. van der Beek and J.A. Mydosh, Supercond. Sci. Technol. 1:242 (1989).

[4] E.V. Thuneberg, in "Proc. Int. Symp. on Flux Pinning and Electromagnetic Properties in Superconductors," T. Matsushita, K. Yamafuji and F. Irie, ed., Fukuoka, Japan, 26 (1985). See also: E.V. Thuneberg, Cryogenics 29:236 (1989).

[5] P.H. Kes, A. Pruymboom, J. van den Berg and J.A. Mydosh, Cryogenics 29:228 (1989).

[6] A. Pruymboom, Thesis, Leiden (1988).

[7] P.G. de Gennes, "Superconductivity of Metals and Alloys," Benjamin, New York (1966).

[8] J.R. Clem, J. Low Temp. Phys. 18:427 (1975).

[9] A.I. Larkin and Yu.N. Ovchinnikov, J. Low Temp. Phys. 34:409 (1979).

[10] P.H. Kes, Physica C 153–155:1121 (1988).

[11] P.H. Kes, J. van den Berg, C.J. van der Beek, J.A. Mydosh, L.W. Roeland, A.A. Menovsky, K. Kadowaki and F.R. de Boer, in: "Proceedings of the First Latin American Conference on High Temperature Superconductivity," R. Nicolsky, R.A. Barrio, O.F. de Lima and R. Escudero, ed., World Scientific, Singapore, 239 (1988).

[12] N.R. Werthamer, E. Helfand and P.C. Hohenberg, Phys. Rev. 147:295 (1966).

[13] G.J. Dolan, G.V. Chandrashekhar, T.R. Dinger, C. Feild and F. Holtzberg, Phys. Rev. Lett. 62:827 (1989).

[14] J.J. Lin, E.L. Benitez, S.J. Poon, M.A. Subramanian and A.W. Sleight, Phys. Rev. B 38:5095 (1988).

SUPERCONDUCTING PHASE DIAGRAM OF A HIGH Tc CERAMIC

F. de la Cruz, L. Civale, and H. Safar

Centro Atómico Bariloche
8400 S.C. de Bariloche
Argentina

INTRODUCTION

In this work we focus our attention on the magnetic response of the superconducting state of high Tc ceramic materials. The superconducting granularity of La-Sr-Cu-O ceramics is controlled by small changes of oxygen concentration. The experimental results show evidence of a hierarchy of weak links connecting superconducting islands.

The granularity is related to superconducting islands surrounded by a non superconducting material. The size of the islands is smaller than the ceramic grains and variations of oxygen concentration of few parts per thousand are sufficient to induce new weak links within the original superconducting islands of the oxygenated material.

Results of heat capacity, electrical resistivity and magnetization of samples with different oxygen concentration are used to provide a phenomenological picture that can be compared to models representing the present theoretical understanding. Many of the experimental results used in the discussion have been already published and references are provided. Experimental details can be found in those references.

The only figure in this paper shows a fairly complete experimental H-T phase diagram of a sample of $La_{1.8}Sr_{0.2}O_{4-\delta}$ where 0.3% of the oxygen has been removed, with reference to the most oxygenated state of the sample. Other phase diagrams, including that of a fully oxygenated sample, will be published elsewhere, but the one presented here has most of the necessary elements to

141

understand the granular superconducting behavior of the ceramics.
To our knowledge this is the first complete phase diagram showing
the behavior of a high Tc granular superconductor.

The paper is organized as follows: after the introduction we
discuss topological configurations of the modulation of the
superconducting order parameter leading to granular behavior.
This is followed by a description of the experimental procedure
used to determine the phase diagram.   Finally, a comparative
discussion of the phase diagram with theoretical models is
provided.

MODULATION OF THE ORDER PARAMETER: CHARACTERISTIC LENGTHS

The granular behavior of the high Tc superconductors was
pointed out by Müller et.al.[1] at the beginning of the study of
the magnetic response of ceramic materials.   The most striking
feature found by those authors was the experimental evidence of a
reversible thermodynamic region in the H-T phase diagram, limited
by the upper critical field curve, $H_{c2}(T)$, and a reasonably well
defined reversibility line[1], $H_R(T)$.

A reversible magnetic behavior implies zero critical current.
In conventional superconductors the critical current is determined
by the pinning of magnetic vortices on lattice defects. In this
sense reversibility implies an almost ideal crystalline structure
of the material.   This is obviously not the case in ceramics and
the reversible magnetization turns out to be one of the intriguing
relevant questions that remains to be fully understood.

At the beginning[1] the reversibility line was assumed to be
the response of a "superconducting glass state" associated[2,3] to
the behavior of multiple connected superconducting networks.

Many other results[4,5,6] using different experimental
techniques have proved that granularity is a superconducting
characteristic of ceramic superconductors.   Nevertheless, a
certain degree of granularity has been shown to be present[7] in
"single crystals" and quite recently, it has been found[8] that the
low Tc ceramic $LiTi_2O_4$ can show an extreme granular behavior,
where superconductivity is concentrated in isolated
superconducting islands occupying near 60% of the sample volume.
These results show that granularity can be determined by a special
topological distribution of the defects and may not be necessarily
related to high Tc superconductivity.   On the other hand, the
reversibility line is present in ceramics and single crystals and,
in a recent experiment, it was shown[9] that the reversibility
line of a $Bi_2Sr_2CaCu_2O_8$ single crystal   coincides with that of the

corresponding ceramic. Although a more systematic work is pending the results of ref. 9 and those reported here open again the question about the origin of the reversibility line.

The glass-like model is not the only one used to describe the magnetic response of the ceramic materials. In a recent review A. Malozemoff[10] has critically discussed the experimental magnetic data in ceramics and single crystals within the framework provided by the glass picture, and a more conventional model[10] where the reversibility line is related to depinning due to thermally induced vortex motion.

Tinkham and Lobb[11] have also discussed the experimental results showing that the granular[12] and the flux pinning[10] models can be considered limiting cases of the same picture.

From the beginning of the research in the area it has been realized[5] that the small coherence length, $\xi(T)$, of the ceramic superconductors was a decisive ingredient to allow the space modulation of the order parameter, $\psi(x)$. As a consequence, there is a tendency to believe[11] that the granular behavior is associated to the depression of the order parameter at grain boundaries. The results in single crystals and in ceramics with different oxygen content show[13] that superconducting granularity should not be confused with the polycrystalline grain boundaries.

It is convenient to emphasize the necessary conditions that the modulation of the order parameter should fulfill to induce a granular behavior in a sample. The order parameter should be depressed in regions that surround superconducting islands, where the order parameter has a larger absolute value. Another necessary condition is that the non superconducting surface regions surrounding the islands (with thickness of the order of $\xi(T)$) extends to distances that are large at the scale of the superconducting penetration depth of the material, $\lambda(T)$. The first condition, when extended to the whole sample, allows superconducting percolation below a temperature, Tc, where a full coherent state is induced among the superconducting islands. The percolation takes place at a lower temperature than the onset temperature, Tco, where superconductivity is nucleated in the islands. The second condition is necessary to modify the thermodynamic properties related to the superconducting currents i.e. the lower critical field, H_{c1}, critical currents, etc..

Up to now the origin of the superconducting islands has not been established. We believe it is not always related to the same type of structural defects. In particular, in this work we emphasize that the granularity can be controlled by the removal of less than one percent of the oxygen content. This is an interesting result, pointing out that oxygen vacancies should be highly correlated in order to induce the necessary topological distribution of the modulation of the order parameter.

PHASE DIAGRAM: EXPERIMENTAL TECHNIQUES

Following the granular description of reference 12 it is possible to point out which are the most relevant features that should be present in a superconducting phase diagram. We discuss these features and the experimental methods used to detect the critical fields in the granular superconductor $La_{1.8}Sr_{0.2}CuO_{4-\delta}$.

At H=O and temperatures between Tco and Tc the thermodynamic state is characterized by a superconducting coherent state within the islands and a lack of coherence among different islands. In this temperature region the thermodynamic properties are dominated by the phase fluctuations of the order parameter between islands. The lack of phase coherence precludes the existence of a full Meissner state and no superconducting current can be established throughout the sample. In this regime the Meissner flux expulsion, at low enough fields, is essentially due to the expulsion from superconducting islands. In this field and temperature range the zero field cooling (ZFC) flux expulsion should be reversible in temperature and proportional to the applied field. At fields higher than the lower critical field of the islands, $H_{c1}^{G}(T)$, vortices will penetrate into them. The vortex state is usually characterized by pinning and, consequently, the flux expulsion should show a non reversible temperature dependence and a lack of field scaling. The description given above clearly suggests how to use flux expulsion measurements to determine the experimental phase diagram.

One of the techniques that has proved to be useful and complementary to determine most of the different critical fields, in ceramics and single crystals, is the measurement of the remanent moment[13]. In this work we define the remanent moment as the residual moment left when removing a magnetic field, applied after the sample was cooled in zero field. The applied field in this context is called inducing field. This experiment, together with the usual ZFC and FC magnetic measurements were used to determine the phase diagram discussed later. In particular, the minimum inducing field necessary to produce a nonzero remanent moment determines $H_{c1}(T)$.

The superconducting coherence throughout the sample is established at temperatures below Tc. As a consequence, a full Meissner effect should be expected below a lower critical field $H_{c1}^{L}(T)$, essentially determined by the characteristic energy coupling the islands.   The search for $H^L_{c1}(T)$ was made measuring the remanent moment and the reversibility of the ZFC expulsion. Above $H^L_{c1}(T)$ flux penetrates in the form of fluxons between the superconducting islands.   If the inducing field exceeds $H^G_{c1}(T)$, flux also penetrates as vortices into the islands.   The measurement of the temperature dependence of the remanent moment provides the possibility to distinguish between the contribution of fluxons and vortices.   If the inducing field has not reached $H^G_{c1}(T)$ the remanent moment disappears at Tc.   On the contrary, if the remanent moment persists up to Tco it indicates that vortices have been induced into the islands.   A clear example of this behavior can be found in ref. 13.

    Another critical field related to granular behavior is $H^L_R(T)$, defined[11,12] as the field where the shielding capability of the currents flowing through the links is strongly reduced.   The determination of this field is more difficult.   However, the measurement of the major magnetization loops around the sample has been proved[14] to be a useful technique to measure the critical current associated to the links.   This is the method used in this work to determine the field at which the links critical currents tend to zero and, consequently, $H^L_R(T)$.

    The reversibility line is defined by the field $H_R(T)$, where the ZFC and FC flux expulsion becomes reversible.

    The ZFC and FC flux expulsion becomes zero at the upper critical field of the grains, $H_{c2}(T)$.   Within the range of fields used in this work $H_{c2}(T)$ can be taken as a vertical line at Tco.

    In the next section we show that deoxygenation of the La-Sr-Cu-O system provides a unique opportunity to control the difference between Tc and Tco.   In this way it is possible to distinguish experimentally the results characteristic of the superconducting islands from those related to the links.

PHASE DIAGRAM: RESULTS AND DISCUSSION

    Figure 1 shows the phase diagram obtained using the experimental techniques described previously.   In this case the phase diagram corresponds to a $La_{1.8}Sr_{0.2}CuO_{4-\delta}$ ceramic with $\delta = 0.012$, as estimated from the loss of weight of the sample heated[15] under vacuum.   The results shown in fig. 1 are representative of the behavior of the ceramics.   Results from the same sample with other oxygen content are also available[16] and will be published elsewhere. They complement and confirm the results and interpretation provided in this paper.

Measurements of remanent moments at 4K indicate that $H^L_{c1}(0)$ is well below 1mOe (the lowest accessible field in our experiments). This is not surprising since the same type of measurements in the fully oxygenated sample gave[16] $H^L_{c1}(0) \simeq 6$mOe. The depression of $H^L_{c1}(T)$ of the deoxygenated sample indicates that oxygen vacancies decrease the critical current of the original links. This result demonstrates that at least part of the oxygen leaves the interconnecting weak link.

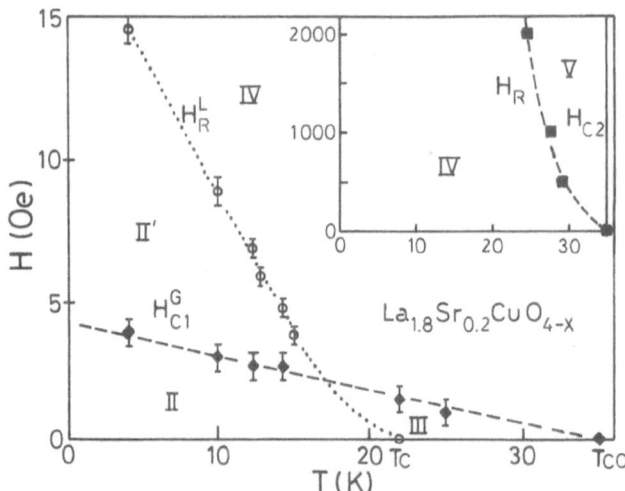

Fig. 1. Superconducting phase diagram of a deoxygenated ceramic sample of La-Sr-Cu-O.

The temperature dependence of the remanent moment was used to determine $H^G_{c1}(T)$, as well as the phase locking temperature Tc. Resistive measurements performed[13] simultaneously in the sample show that the resistivity goes to zero at Tc, as expected from the previous discussion. The ZFC and FC flux expulsion goes to zero at Tco, where the electrical resistivity starts to drop from its normal value.   At this temperature the superconducting phase coherence is established only within the islands, while a Tc superconductivity is phase locked among islands.

The field where the links critical current is quenched, $H_R^L(T)$, is shown in the figure.   In this oxygenation state of the sample $H_R^L(T)$ intersect $H_{c1}^G(T)$, generating new magnetic regions in the phase diagram.

We discuss now the magnetic features of the phase diagram shown in fig. 1.   As it has already been discussed, the Meissner state in the whole sample has not been detected down to the lowest accessible fields in these experiments.   The magnetic properties of region II, are characterized by superconducting islands in the Meissner state and magnetic fluxons penetrating between islands. In the region II' the magnetic flux is a result of the superposition of vortices within the islands and fluxons between them.   Above $H_R^L(T)$ the flux expulsion is only due to superconductivity in the islands.   In region III the islands are in the Meissner state, while in IV vortices penetrate into them. In the scale of the diagram, $H_{c2}(T)$ can be taken as a vertical line at Tco.

The results reported here show quite a rich magnetic phase diagram.   The diagram itself is not in contradiction with a glass model neither does it prove its validity.   Nevertheless, it points towards predictions that should be experimentally verified. Before proceeding with some possible suggestions let us point out several experimental results providing new insight to the magnetic behavior of the high Tc materials.

The lower field values of the $H_R^L(T)$ line separates the reversible region III from the non reversible region II.   From this point of view it can be considered a reversibility line. However, at higher fields the same line separatesthe non reversible regions II' and IV, marking a well defined difference with the reversible transition found in oxygenated materials. Nevertheless, ZFC and FC flux expulsion measurements show the existence of another line, $H_R(T)$ separating the irreversible region IV from a reversible region V, as shown in the inset.

A difference between the two reversible regions is that region III corresponds to the Meissner state of isolated islands, while in region V magnetic flux penetrates into the islands.   An important result is that $H_R(T)$ of the deoxygenated material coincides with that of the fully oxygenated state[16].

Up to now the results could be understood assuming that the Josephson links among the superconducting islands of the oxygenated material are weakened by deoxygenation, leaving unperturbed the superconducting islands. In this picture, the reversibility line is related to the internal properties of the islands. On the other hand, the $H_{c1}^G(T)$ of fig. 1 makes evident that deoxygenation modifies the internal structure of the original superconducting islands. The phase diagram of the oxygenated sample[16] and previous experimental results[5] show that $H_{c1}^G(0)=100$ Oe in the oxygenated material. Although a microscopic theory of the high Tc superconductors is still missing, it is reasonable to believe that $H_{c1}(T)$ is still determined by the weak field penetration depth. We found no reasonable argument indicating that a change of 0.3% of oxygen concentration could change the penetration depth in factors of the order of five. The results reported here, together with those for other oxygen concentrations[16], indicate that in the process of deoxygenation part of the oxygen leaves the superconducting islands in a highly correlated manner. The oxygen vacancies are responsible for the creation of new two dimensional surfaces, introducing new internal weak links, as discussed previously. This picture has been made evident by experiments[8] in the low Tc ceramic superconductor $LiTi_2O_4$, indicating that the correlation among defects is not related to the high Tc superconducting properties but is a typical response of the metallic oxides.

It is interesting to see that this picture of highly correlated oxygen vacancies distributed in a rather particular topology reconciles what seems to be contradictory results: the extreme sensitivity of $H_{c1}^G(T)$ to oxygen vacancies on one side and the independence of $H_R(T)$ on oxygen concentration on the other.

Let us now recapitulate the most important results found in this research: the granular picture of the ceramic superconductors is clearly demonstrated, oxygen depletion is shown to be an excellent tool to control the degree of granularity, varying the strength of the original weak links and introducing new links within the original superconducting islands and, finally, it is surprising to show that the fundamental changes introduced in the magnetic response of the material leaves unmodified the reversibility line of the superconducting islands. This and recent results in single crystals[9] bring up again the question about the origin of the reversibility line in the High Tc superconductors.

REFERENCES

1.   K.A. Müller, M. Takashige, and J.G. Bednorz, Phys.Rev.Lett. 58:1143 (1987).
2.   I. Morgenstern, K.A. Müller, and J.G. Bednorz, Z.Phys. B69:33 (1987).
3.   J. Choi and J. José, Phys.Rev.Lett. 62:320 (1989).
4.   D.A. Esparza, C.A. D'Ovidio, J. Guimpel, E. Osquiguil, L. Civale, and F. de la Cruz, Solid State Commun. 63:137 (1987).
5.   F. de la Cruz, L. Civale, and H. Safar, Progress in High Temperature Superconductivity 9:5 (1988), ed., R. Nicolsky, R.A. Barrio, O Ferreira de Lima, and R. Escudero, (Proc. First Latin American Conference on High Temperature Superconductivity, World Scientific Pub. Co., Singapore).
6.   W.K. Kwok, G.W. Crabtree, D.G. Hinks, D.W. Capone, J.D. Jorgensen, and K. Zhang, Phys.Rev. B35:5343 (1987).
7.   H. Küpfer, I. Apfelstedt, R. Flükiger, R. Meyer-Hirmer, W. Schauer, T. Wolf and H. Wühl, Physica C153-155:367 (1988).
8.   L. Civale, H. Pastoriza, and F. de la Cruz, to be published in Solid State Commun.
9.   H. Safar, C. Durán, J. Guimpel, L. Civale, J. Luzuriaga, E. Rodriguez, F. de la Cruz, C. Fainstein, L.F. Schneemeyer, and J.V. Waszczak, to be published in Physical Review B.
10. A. Malozemoff, to be published in Physical Properties of High Temperature Superconductors, ed. D.M. Ginsberg (World Scientific Publ. Co., Singapore, 1989).
11. M. Tinkham and C.J. Lobb, to be published in Solid State Physics, Vol. 42, ed. H. Ehrenreich and D. Turnbull (Academic Press, Orlando, Fl., 1989).
12. J.R. Clem, Physica C153-155:50 (1988); A. Raboutou, P. Peyral, J. Rosenblatt, C. Lebeau, O. Peña, A. Perrin, and M. Sergent, Europhysics Lett. 4:1321 (1987).
13. L. Civale, H. Safar, and F. de la Cruz, Mod.Phys.Lett. B3:173 (1989).
14. M. Oussena, S. Senoussi, and G. Collin, Europhysics Lett. 4:625 (1987).
15. E.J. Osquiguil, R. Decca, G. Nieva, L. Civale, and F. de la Cruz, Solid State Commun. 65:491 (1988).
16. L. Civale, PhD Thesis, Instituto Balseiro, Universidad Nacional de Cuyo, 1989.

MAGNETIC ORDERING OF RARE EARTH (R) IONS IN

$RBa_2Cu_3O_{7-\delta}$ COMPOUNDS (R = Nd, Sm, Dy, Er; $0 \leq \delta \leq 1$)

B. W. Lee, J. M. Ferreira,[*] S. Ghamaty, K. N. Yang,[+]
and M. B. Maple

Department of Physics and Institute for Pure and Applied
Physical Sciences
University of California, San Diego
La Jolla, CA 92093 USA

ABSTRACT

Low temperature specific heat measurements have been performed between ~0.5 K and 30 K on $RBa_2Cu_3O_{7-\delta}$ (R = Nd, Sm, Dy, Er; $0 \leq \delta \leq 1$) compounds in magnetic fields up to 5 kOe. Specific heat anomalies due to antiferromagnetic ordering of R^{3+} ions in the $T_c \approx 92$ K superconducting $RBa_2Cu_3O_{7-\delta}$ ($\delta \approx 0.1$) compounds can be well described by the anisotropic 2-D Ising model with an exchange interaction parameter ratio that ranges from ~50 for R = Nd to ~4 for R = Dy. The magnetic specific heat anomalies and magnetic ordering temperatures of the non-superconducting $RBa_2Cu_3O_{7-\delta}$ ($\delta \gtrsim 0.5$) compounds are markedly different than those of the superconducting compounds, suggesting that Ruderman-Kittel-Kasuya-Yosida (RKKY) or superexchange interactions, as well as dipolar interactions, are involved in the magnetic ordering of the R^{3+} ions in these compounds. The changes in the magnetic specific heat anomalies with oxygen concentration indicate that the oxygen vacancy concentration affects the electronic structure of the $RBa_2Cu_3O_{7-\delta}$ compounds. The magnetic ordering temperatures of the $T_c \approx 92$ K superconducting $RBa_2Cu_3O_{7-\delta}$ compounds, at least for R = Nd, Dy, and Er, move to lower temperatures upon application of an external magnetic field, consistent with the antiferromagnetic ordering of the R^{3+} magnetic moments.

---

*Permanent address: Departamento de Fisica, Universidade Federal de Pernambuco, 5000 Recife, Brazil
+Present address: Ampex Corporation, Audio-Video Systems Division, 401 Broadway, Redwood City, CA 94063

INTRODUCTION

Following the discovery of superconductivity with a superconducting transition temperature $T_c \approx 92$ K in $YBa_2Cu_3O_{7-\delta}$,[1] the series of rare earth (R) barium copper oxide compounds $RBa_2Cu_3O_{7-\delta}$ were also found to exhibit superconductivity with $T_c \approx 92$ K, except for R = Ce, Pr, Pm, and Tb.[2] The superconducting $RBa_2Cu_3O_{7-\delta}$ compounds are isomorphic to the prototypical $YBa_2Cu_3O_{7-\delta}$ compound which has an orthorhombic oxygen-deficient perovskite-type crystal structure. The lattice parameters and unit cell volumes decrease systematically from R = La to Lu.[3] Magnetic susceptibility measurements on the superconducting $RBa_2Cu_3O_{7-\delta}$ compounds in the normal state above T_c revealed that the effective magnetic moments of the R^{3+} ions are close to the free ion values.[4] The generally negative values of the Curie-Weiss temperature are consistent with antiferromagnetic interactions between the R^{3+} ions, as well as modifications to $\chi(T)$ arising from partial lifting of the 2J+1 degeneracy of the Hund's rules ground state multiplets by the crystalline electric field (CEF).

As part of a detailed investigation of the series of $RBa_2Cu_3O_{7-\delta}$ compounds, we have performed low temperature specific heat measurements on these materials between ~0.5 K and 50 K.[5,6] These compounds exhibit a variety of interesting effects that are associated with the ground state 4f electron configuration of the R^{3+} ions. The $RBa_2Cu_3O_{7-\delta}$ compounds with R = Ho, Tm, and Yb display electronic Schottky anomalies due to the splitting of the Hund's rules ground state multiplets of the R^{3+} ions by the CEF.[5] Magnetic ordering in $HoBa_2Cu_3O_{7-\delta}$ at ~0.17 K has been observed by means of specific heat measurements,[7,8] while the onset of hyperfine splitting in $YbBa_2Cu_3O_{7-\delta}$ yields a magnetic ordering temperature of ~0.35 K, according to Mössbauer effect measurements.[9] In $RBa_2Cu_3O_{7-\delta}$ compounds with R = Nd, Sm, Gd, Dy, and Er, the R^{3+} ions order antiferromagnetically with Néel temperatures T_N that range from 0.52 K for R = Nd to 2.25 K for R = Gd. Coexistence of antiferromagnetism and superconductivity in $RBa_2Cu_3O_{7-\delta}$ compounds below T_N has been established by means of specific heat[10] and neutron scattering measurements.[11-15]

Since the R-R distance along the c-axis is about three times larger than the R-R distance in the a-b plane in the $RBa_2Cu_3O_{7-\delta}$ compounds, magnetic ordering of the R^{3+} ions with strong 2-D character could be anticipated in these materials. The values of the entropy for the superconducting $RBa_2Cu_3O_{7-\delta}$ compounds with R = Nd, Sm, Dy, and Er at 5 K, which is well above T_N, are close to $\mathcal{R}\ln 2$,[6] indicating that these compounds have a doublet ground state in the CEF. The non-superconducting materials can be obtained from the superconducting $RBa_2Cu_3O_{7-\delta}$ compounds by increasing the oxygen vacancy concentration δ to values $\gtrsim 0.5$, which is accompanied by a change of the crystal symmetry from orthorhombic to tetragonal.[16] The magnetic susceptibility of non-superconducting $RBa_2Cu_3O_{7-\delta}$ compounds has been investigated and found to be comparable to that of their superconducting counterparts.[17-20] The magnetic susceptibility $\chi(T)$ of the non-superconducting compounds can be described by a Curie-Weiss law with an effective magnetic moment that is close to the free ion value. Specific heat measurements on superconducting and non-superconducting $GdBa_2Cu_3O_{7-\delta}$ compounds[21,22] yielded no significant change in T_N nor shape of the specific heat anomaly associated with the antiferromagnetic ordering. However, specific heat measurements on non-superconducting $RBa_2Cu_3O_{7-\delta}$ compounds with R = Nd and Sm revealed that the magnetic ordering temperatures and shapes of the magnetic specific heat anomalies in these compounds are dramatically different than those of the superconducting $RBa_2Cu_3O_{7-\delta}$ compounds.[15,23] Nakazawa et al.[24] reported two different types of magnetic phase transitions in specific heat measurements on $DyBa_2Cu_3O_{7-\delta}$ compounds.

In this paper, we demonstrate that the specific heat anomalies due to magnetic ordering can be described in terms of an anisotropic 2-D Ising model for the $T_c \approx 92$ K superconducting $RBa_2Cu_3O_{7-\delta}$ ($\delta \approx 0.1$) compounds (R = Nd, Sm, Dy, Er), and document the changes in the magnetic specific heat anomalies that occur with increasing oxygen vacancy concentration δ for R = Nd, Dy, and Er, in magnetic fields up to 5 kOe.

EXPERIMENTAL DETAILS

The $T_c \approx 92$ K superconducting $RBa_2Cu_3O_{7-\delta}$ samples were fabricated by means of standard solid state reaction methods.[25] The $NdBa_2Cu_3O_{7-\delta}$ compounds with different oxygen vacancy concentrations δ were prepared by annealing $T_c \approx 92$ K superconducting $NdBa_2Cu_3O_{7-\delta}$ compounds at appropriate temperatures in flowing Ar gas; sample preparation procedures and other experimental details will be reported elsewhere.[26] Pieces of $RBa_2Cu_3O_{7-\delta}$ samples with R = Dy and Er that were used in our previously reported specific heat measurements[6] were quenched after annealing in air at 860°C for 22 hours in order to reduce the oxygen concentration. In the case of $DyBa_2Cu_3O_{7-\delta}$, this suppressed the superconductivity and converted the material to the tetragonal structure, while in the case of $ErBa_2Cu_3O_{7-\delta}$, the sample remained superconducting with $T_c \approx 70$ K and was not transformed completely to the teragonal structure. In order to obtain a completely non-superconducting $ErBa_2Cu_3O_{7-\delta}$ sample, another piece of $ErBa_2Cu_3O_{7-\delta}$, that had also been quenched from 860°C in air, was annealed at 800°C in flowing He gas and furnace cooled to room temperature.

Oxygen concentrations in the $RBa_2Cu_3O_{7-\delta}$ compounds were determined by iodometric titration and by the change in weight before and after reaction steps. The critical concentration of oxygen at which superconductivity is suppressed appears to vary for different $RBa_2Cu_3O_{7-\delta}$ compounds prepared by identical methods and is presently being investigated. Powder X-ray diffraction, electrical resistivity, and magnetization measurements were used to characterize the $RBa_2Cu_3O_{7-\delta}$ compounds with different oxygen concentrations. The $T_c \approx 92$ K superconducting $RBa_2Cu_3O_{7-\delta}$ compounds were all found to be isostructural with the orthorhombic $T_c \approx 92$ K superconducting $YBa_2Cu_3O_{7-\delta}$ prototype, and the non-superconducting $RBa_2Cu_3O_{7-\delta}$ compounds were all determined to have the tetragonal stucture characteristic of non-superconducting $YBa_2Cu_3O_{7-\delta}$ with $\delta \gtrsim 0.5$. Electrical resistivity measurements were made using a low frequency (17 Hz) ac four-lead technique and dc magnetic susceptibility measurements were performed with a variable temperature SQUID susceptometer (SHE Corp.). A ^{3}He semi-adiabatic calorimeter using a standard heat pulse technique was employed for the specific heat measurements at temperatures between ~ 0.5 K and 30 K. Electrical resistivity,[2] magnetization,[4,25] and specific heat[6] measurements on the superconducting

$RBa_2Cu_3O_{7-\delta}$ samples from which the non-superconducting $RBa_2Cu_3O_{7-\delta}$ compounds were prepared as described above have been reported elsewhere.

RESULTS AND DISCUSSION

Shown in Fig. 1 are magnetic specific heat ΔC versus temperature data between ~0.5 K and 3 K for the $T_c \approx 92$ K $RBa_2Cu_3O_{7-\delta}$ compounds ($\delta \approx 0.1$) with R = Nd, Sm, Dy, Er. The $\Delta C(T)$ data were obtained by subtracting $C(T)$ data for the $T_c \approx 92$ K superconducting $YBa_2Cu_3O_{7-\delta}$ compound[5] from $C(T)$ data for the $T_c \approx 92$ K superconducting $RBa_2Cu_3O_{7-\delta}$ compounds,[6] in order to remove the conduction electron and lattice contributions. The method of least squares has been used to fit the data for $0.5 \lesssim T \leq 3$ K ($0.5 \lesssim T \leq 4$ K for $NdBa_2Cu_3O_{7-\delta}$) with the Onsager solution of the anisotropic 2-D Ising model, as described in reference 15; the resultant fits, which are represented by the solid lines in Fig. 1, yield the Néel temperature T_N and the exchange interaction parameters E_1 and E_2 within the a-b plane of the orthorhombic $RBa_2Cu_3O_{7-\delta}$ crystal structure. The exchange interaction parameter ratio E_1/E_2 ranges from ~50 for R = Nd to ~4 for R = Dy. The values of T_N, E_1 and E_2 for the $RBa_2Cu_3O_{7-\delta}$ compounds are listed in Table 1.

The Ising model analysis of the magnetic specific heat $\Delta C(T)$ data for the $T_c \approx 92$ K superconducting $RBa_2Cu_3O_{7-\delta}$ compounds suggests that the exchange interactions within the a-b plane are anisotropic and much stronger than the interaction between the planes. The source of anisotropy within the a-b plane is not readily apparent, since these materials have a rather small orthorhombic distortion and are nearly tetragonal. In fact, the lattice parameter ratio within the a-b plane, a/b, which describes the orthorhombic distortion of the $RBa_2Cu_3O_{7-\delta}$ ($\delta \approx 0.1$) compounds, increases as R moves toward the heavier rare earths, in contrast to the exchange interaction parameter anisotropy ratio, E_1/E_2, which decreases from the

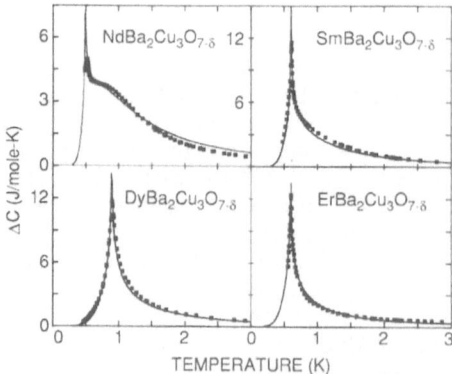

Figure 1. Magnetic specific heat ΔC, corrected for electronic and lattice contributions, versus temperature T for superconducting $RBa_2Cu_3O_{7-\delta}$ ($\delta \approx 0.1$) compounds with R = Nd, Sm, Dy, and Er. The solid lines represent fits of the anisotropic 2-D Ising model to the $\Delta C(T)$ data as described in the text.

Table 1.   The exchange interaction parameters E_1 and E_2 within the a-b plane of the $RBa_2Cu_3O_{7-\delta}$ crystal structure and the experimental (calculated) Néel temperatures T_N from fits of the anisotropic Ising model to the specific heat data.

$RBa_2Cu_3O_{7-\delta}$	E_1 (K)	E_2 (K)	E_1/E_2	T_N (K)
Nd	0.85	0.017	50	0.52 (0.50)
Sm	0.70	0.063	11	0.61
Dy	0.73	0.18	4	0.90
Er	0.54	0.10	5	0.60

lighter to the heavier rare earths. If the magnetic interactions between the R^{3+} magnetic moments were mediated by electrons, via RKKY or superexchange interactions, E_1 and E_2 would be expected to be comparable to one another. A possible source of the exchange interaction anisotropy could be the overlap between the R^{3+} 4f electron CEF ground state wave functions and valence band states of the CuO_2 planes whose carrier concentration is governed by the concentration of oxygen vacancies in the CuO chains.

Displayed in Fig. 2 are specific heat C versus temperature data for $NdBa_2Cu_3O_{7-\delta}$ compounds with different oxygen concentrations. Unlike the superconducting and non-superconducting $GdBa_2Cu_3O_{7-\delta}$ compounds, in which the shapes of the magnetic specific anomaly and the magnetic ordering temperatures are virtually identical,[21,22] there is a dramatic change in the magnetic specific heat anomaly with decreasing oxygen concentration. The C(T) data for the $NdBa_2Cu_3O_{7-\delta}$ ($\delta \approx 0.1$) compound resemble our previously reported C(T) data for another $NdBa_2Cu_3O_{7-\delta}$ compound[6] (see Fig. 1), with a somewhat narrower

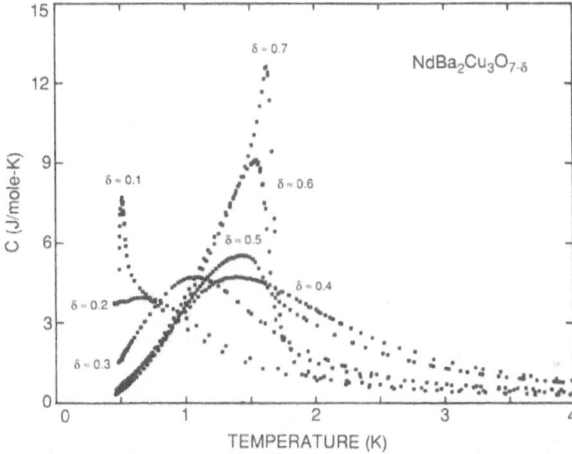

Figure 2. Specific heat C versus temperature T data for 0.5 K $\leq$ T $\leq$ 4 K of $NdBa_2Cu_3O_{7-\delta}$ compounds with values of δ in the range $0.1 \leq \delta \leq 0.7$.

rounded feature above the Néel temperature $T_N = 0.51$ K and a higher sharp peak at T_N. Neutron scattering measurements on a $NdBa_2Cu_3O_{6.9}$ compound in conjunction with our previous specific heat studies[15] indicated 3-D antiferromagnetic ordering of the Nd^{3+} moments along the c-axis. Although the specific heat curve in Fig. 2 for the $NdBa_2Cu_3O_{7-\delta}$ compound with $\delta \approx 0.1$ is associated with antiferromagnetic order, only a broad rounded specific anomaly is found for the $NdBa_2Cu_3O_{7-\delta}$ compound with $\delta \approx 0.2$. It is quite surprising that such a small difference in oxygen concentration results in the suppression of the sharp peak in the $NdBa_2Cu_3O_{7-\delta}$ ($\delta \approx 0.2$) compound, even though the $\delta \approx 0.1$ and 0.2 compounds both have the orthorhombic crystal structure and exhibit superconductivity.

Magnetic ordering in the $NdBa_2Cu_3O_{7-\delta}$ system may be suppressed at temperatures above ~0.5 K for oxygen vacancy concentrations δ in the range $0.2 \lesssim \delta \lesssim 0.4$. Such a suppression of magnetic order could be caused by disorder within the CuO_2 planes if the Nd^{3+} moments are coupled via the RKKY or superexchange interactions. For $\delta \gtrsim 0.5$, where $NdBa_2Cu_3O_{7-\delta}$ loses its superconductivity, magnetic ordering reappears as the oxygen vacancy concentration δ is increased. Magnetization data on another non-superconducting $NdBa_2Cu_3O_{7-\delta}$ ($\delta \approx 0.5$) compound indicated that the Nd^{3+} ions ordered antiferromagnetically, while neutron scattering measurements on a non-superconducting $NdBa_2Cu_3O_{6.3}$ compound revealed the same type of 3-D antiferromagnetic ordering of the Nd^{3+} moments as had been found for the superconducting $NdBa_2Cu_3O_{6.9}$ compound.[15] The reappearance of the antiferromagetic order at higher oxygen vacancy concentrations $\delta \gtrsim 0.5$ could be induced by the antiferromagnetic ordering of the Cu^{2+} ions which occurs for $\delta \gtrsim 0.5$, after superconductivity has been quenched.

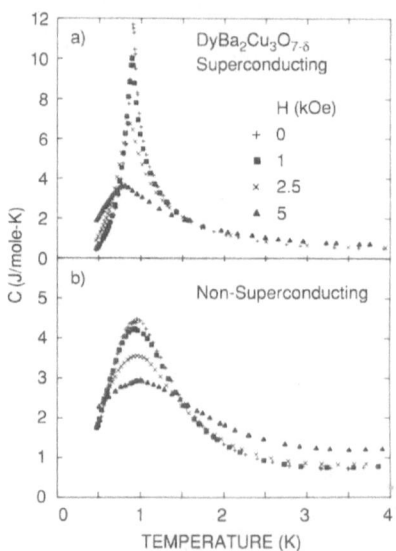

Figure 3. Specific heat C versus temperature data for a) superconducting ($\delta \approx 0.1$) and b) non-superconducting ($\delta \approx 0.6$) $DyBa_2Cu_3O_{7-\delta}$ compounds in magnetic fields H = 0, 1, 2.5, and 5 kOe.

Shown in Fig. 3 are specific heat C versus temperature data for $0.5 \text{ K} \lesssim \text{T} \leq 4$ K for superconducting and non-superconducting $DyBa_2Cu_3O_{7-\delta}$ compounds in magnetic fields H up to 5 kOe. The data for the superconducting $DyBa_2Cu_3O_{7-\delta}$ compound have been reported previously.[6] The shape of specific heat anomaly due to magnetic ordering in the superconducting $DyBa_2Cu_3O_{7-\delta}$ ($\delta \approx 0.1$) compound is quite different from that of the non-superconducting $DyBa_2Cu_3O_{7-\delta}$ ($\delta \approx 0.6$) compound; C(T) exhibits a sharp peak at $T_N = 0.90$ K in the superconducting compound and a more rounded maximum, also at about 0.9 K, in the non-superconducting compound. In the superconducting compound, T_N decreases with applied magnetic field, consistent with the antiferromagnetic ordering of the Dy^{3+} magnetic moments. However, in the non-superconducting compound, the maximum in C(T) moves to higher temperatures as H is increased, suggesting that the interaction between the Dy^{3+} ions in the non-superconducting compound may be

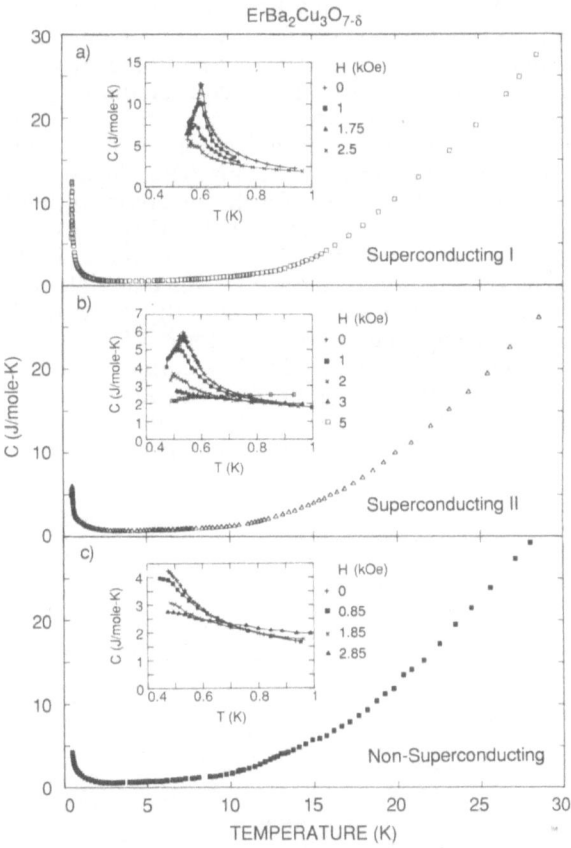

Figure 4. Specific heat C versus temperature T data for $0.5 \text{ K} \lesssim \text{T} \leq 30$ K for three different $ErBa_2Cu_3O_{7-\delta}$ compounds; superconducting I ($\delta \approx 0.1$; open squares), superconducting II ($\delta \approx 0.5$; open triangles), and non-superconducting ($\delta \approx 0.7$; solid squares). Insets: C versus T in various magnetic fields.

different from that in the superconducting compound. Neutron scattering studies[13,14] on the superconducting $DyBa_2Cu_3O_{7-\delta}$ compound have shown that the Dy^{3+} magnetic moments are aligned antiferromagnetically along the c-axis. According to magnetization measurements on the non-superconducting $DyBa_2Cu_3O_{7-\delta}$ compound,[18,19,20] an antiferromagnetic transition occurs at ~0.9 K, close to the temperature at which C(T) exhibits a maximum.

Specific heat C versus temperature data in the range $0.5 \text{ K} \lesssim T \leq 30 \text{ K}$ are shown in Fig. 4 for three different $ErBa_2Cu_3O_{7-\delta}$ compounds; a) superconducting I with $T_c = 92 \text{ K}$, $\Delta T_c = 2 \text{ K}$ ($\delta \approx 0.1$), b) superconducting II with $T_c = 72 \text{ K}$, $\Delta T_c = 18 \text{ K}$ ($\delta \approx 0.5$), and c) non-superconducting ($\delta \approx 0.7$). The results for the superconducting I $ErBa_2Cu_3O_{7-\delta}$ compound with $T_c = 92 \text{ K}$ are represented by open squares in Fig. 4, and have been reported previously.[6] The general shapes of the specific heat curves are quite similar to each other, except that the C(T) data above ~5 K for the non-superconducting $ErBa_2Cu_3O_{7-\delta}$ compound are somewhat higher than those of the superconducting $ErBa_2Cu_3O_{7-\delta}$ compounds. This indicates that the splitting between the Er^{3+} 4f electron CEF-split energy levels is smaller for the non-superconducting compound. The shape of the magnetic specific heat anomaly and the magnetic ordering temperature in the superconducting $ErBa_2Cu_3O_{7-\delta}$ compounds are quite different from those of the non-superconducting $ErBa_2Cu_3O_{7-\delta}$ compound. The superconducting $ErBa_2Cu_3O_{7-\delta}$ compounds exhibit specific heat peaks at 0.60 K and 0.53 K for the superconducting I and II samples, respectively. For the non-superconducting $ErBa_2Cu_3O_{7-\delta}$ compound, we were unable to observe a peak in C(T) above ~0.5 K, the lower limit of ^{3}He calorimeter, indicating that magnetic ordering, if it exists in this material, occurs below 0.5 K. Recently, Simizu et al.[27] reported that nominally tetragonal $ErBa_2Cu_3O_{7-\delta}$ compounds did not show any sharp anomalies for $\delta \approx 0.66 - 0.88$. The C(T) data for the superconducting I and II $ErBa_2Cu_3O_{7-\delta}$ compounds in various magnetic fields up to 5 kOe displayed in the inset of Figs. 4a and 4b reveal that T_N moves to lower temperature as H is increased, consistent with the antiferromagnetic ordering of the Er^{3+} ions established with neutron scattering measurements.[12]

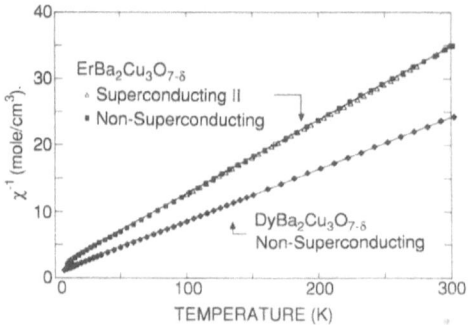

Figure 5. Inverse magnetic susceptibility χ^{-1} in the normal state, measured in a magnetic field of 990 Oe, versus temperature for superconducting II (triangles) and non-superconducting (solid squares) $ErBa_2Cu_3O_{7-\delta}$ and non-superconducting $DyBa_2Cu_3O_{7-\delta}$ (solid diamonds). The solid lines represent fits of Eq. (1) to the data as described in the text.

Table 2.   Effective magnetic moment μ_{eff} and Curie-Wiess temperature θ
for $RBa_2Cu_3O_{7-\delta}$ compounds determined from fits of Eq. (1) to
the $\chi(T)$ data above 100 K.

	μ_{eff} (μ_B)	θ (K)
non-superconducting $DyBa_2Cu_3O_{7-\delta}$	10.2	-11.3
superconducting II $ErBa_2Cu_3O_{7-\delta}$	8.5	-13.3
non-superconducting $ErBa_2Cu_3O_{7-\delta}$	8.5	-15.5

Inverse magnetic susceptibility χ^{-1} versus temperature data, from $\chi(T)$ measurements in an applied field of 990 Oe, are shown in Figure 5. Solid diamonds represent data for the non-superconducting $DyBa_2Cu_3O_{7-\delta}$ compound, and open triangles and solid squares depict data for the superconducting II and non-superconducting $ErBa_2Cu_3O_{7-\delta}$ compounds, respectively. The lines in Fig. 5 represent fits of the $\chi(T)$ data for the non-superconducting compounds above 100 K with a Curie-Weiss law; i.e.,

$$\chi(T) = N\mu_{eff}^2/3k_B(T - \theta) \tag{1}$$

where N is Avogadro's number, μ_{eff} is the effective magnetic moment, and θ is the Curie-Weiss temperature. Below 100 K, the $\chi^{-1}(T)$ data fall below the linear Curie-Weiss curves. Values of μ_{eff} and θ obtained from fits of Eq. (1) to the $\chi(T)$ data are listed in Table 2. The effective magnetic moments are comparable to the R^{3+} free ion values. The negative values of the Curie-Weiss temperatures are consistent with antiferromagnetic interactions between the R^{3+} magnetic moments, but may also reflect modifications to $\chi(T)$ due to CEF splitting of the R^{3+} 4f electron energy levels. Specific heat data for superconducting $RBa_2Cu_3O_{7-\delta}$ compounds indicate that the ground and first excited states are doublets split by 40 K for R = Dy and 90 K for R = Er.[7]

CONCLUDING REMARKS

The conduction electron partial density of states at the Fermi level at the R sites appears to be extremely low in superconducting $RBa_2Cu_3O_{7-\delta}$ compounds.[28,29,30] This provides an explanation for the rather weak interaction between the superconducting electrons and the R magnetic moments which, in turn, allows superconductivity and antiferromagnetic ordering of the R ions to coexist in the $RBa_2Cu_3O_{7-\delta}$ compounds with R = Nd, Sm, Gd, Dy, and Er.[6] As discussed above, low temperature specific heat measurements[21,22] on $GdBa_2Cu_3O_{7-\delta}$ have shown that the feature in C(T) due to antiferromagnetic ordering of the Gd^{3+} ions and the Néel temperature T_N are identical for both superconducting and non-superconducting compounds. However, in addition to the low temperature specific heat data presented herein, our previous specific heat studies[15,23] have revealed striking differences in the magnetic contributions to C(T) between superconducting and non-superconducting $RBa_2Cu_3O_{7-\delta}$ compounds with R = Nd and Sm.[15,23] We found that the magnetic specific heat of the $T_c \approx 92$ K superconducting $NdBa_2Cu_3O_{7-\delta}$ compound consists of a sharp peak at 0.52 K superimposed on a rounded feature that can be described by the anisotropic 2-D Ising model with an exchange interaction ratio $E_1/E_2 \approx 50$. However, the magnetic specific heat anomaly of the non-superconducting $NdBa_2Cu_3O_{7-\delta}$ compound has a markedly different shape than that of the superconducting $NdBa_2Cu_3O_{7-\delta}$ compound, with a large enhancement of the magnetic ordering temperature, consistent with the more complete results presented herein. We also observed that the magnetic specific heat anomaly of the $T_c \approx 92$ K superconducting $SmBa_2Cu_3O_{7-\delta}$ compound can be well described by the anisotropic 2-D Ising model with $T_N = 0.61$ K and $E_1/E_2 \approx 11$. In contrast, the magnetic specific heat anomaly of the non-superconducting $SmBa_2Cu_3O_{7-\delta}$ compound has a rounded shape with a maximum at ~1 K and can be accounted for either by a 1-D Ising chain or by a Schottky anomaly for a two level system split by 2.4 K. Since Gd is an S-state ion with orbital angular momentum $L = 0$, it is not subject to the CEF, in the first approximation. However, the Nd, Sm, Dy, and Er ions have $L = 6, 5, 5$, and 6, respectively, and are strongly affected by the CEF. The CEF at the sites of the R ions influences the magnetic properties of the $RBa_2Cu_3O_{7-\delta}$ compounds; variations in the CEF associated with structural changes could account, at least in part, for differences in the magnetic properties of superconducting and non-superconducting $RBa_2Cu_3O_{7-\delta}$ compounds.

The lack of conduction electron density around the R sites suggests that the mechanism for the low temperature magnetically ordered states in the superconducting $RBa_2Cu_3O_{7-\delta}$ compounds may not be the RKKY interaction, which is mediated by conduction electrons, but could instead be a modified RKKY interaction[31] or the combined effects of dipolar interactions coupled with anisotropy associated with the CEF and superexchange via covalent bonds between the R^{3+} ions and neighboring oxygen ions. Evidence in favor of the RKKY interaction is the observed scaling of the Néel temperatures T_N of the $T_c \approx 92$ K superconducting $RBa_2Cu_3O_{7-\delta}$ compounds from the value for Gd according to the deGennes factor $(g_J - 1)^2 J(J+1)$, where g_J and J are, respectively, the Landé g-factor and total angular momentum of the R^{3+} ion.[6,32]

The suppression of magnetic ordering in the superconducting $NdBa_2Cu_3O_{7-\delta}$ compounds with oxygen concentration δ between ~0.2 and ~0.4 may be due to disorder within the CuO_2 planes. Magnetic ordering of non-superconducting $NdBa_2Cu_3O_{7-\delta}$ compounds with $\delta \gtrsim 0.5$ appears to be induced by the occurrence of antiferromagnetic ordering of the Cu^{2+} ions. The apparent absence of magnetic ordering of the Er^{3+} ions in the non-superconducting $ErBa_2Cu_3O_{7-\delta}$ compounds may be associated with the proclivity of the Er^{3+} moments to lie in the a-b plane.

ACKNOWLEDGMENTS

We would like to thank E. A. Early, J. T. Markert, and C. L. Seaman for their assistance. This research was supported by the U. S. Department of Energy under Grant No. DE-FG03-86ER45230. Support from the U. S. National Science Foundation—Low Temperature Physics—Grant No. DMR84-11839 (BWL) and Conselho Nacional de Desenvolvimento Científico e Technológico (CNPq), Coordenação de Aperfeiçoamento de Pessoal do Ensino Superior (CAPES), and

Financiadora de Estudos e Projetos (FINEP) of Brazil (JMF) is gratefully acknowledged.

REFERENCES

1. M. K. Wu, J. R. Ashburn, C. J. Torng, P. H. Hor, R. L. Meng, L. Gao, Z. J. Huang, Y. Q. Wang, and C. W. Chu, Phys. Rev. Lett. 58:908 (1987).
2. See, for example, M. B. Maple, Y. Dalichaouch, J. M. Ferreira, R. R. Hake, B. W. Lee, J. J. Neumeier, M. S. Torikachvili, K. N.Yang, H. Zhou, R. P. Guertin, and M. V. Kuric, Physica B 148:155 (1987), and reference to other work cited therein.
3. K. N. Yang, Y. Dalichaouch, J. M. Ferreira, R. R. Hake, B. W. Lee, J. J. Neumeier, M. S. Torikachvili, H. Zhou, and M. B. Maple, Jpn. J. Appl. Phys. 26-3:1037 (1987).
4. H. Zhou, C. L. Seaman, Y. Dalichaouch, B. W. Lee, K. N. Yang, R. R. Hake, M. B. Maple, R. P. Guertin, and M. V. Kuric, Physica C 152:321 (1988).
5. J. M. Ferreira, B. W. Lee, Y. Dalichaouch, M. S. Torikachvili, K. N. Yang, and M. B. Maple, Phys. Rev. B 37:1580 (1988).
6. B. W. Lee, J. M. Ferreira, Y. Dalichaouch, M. S. Torikachvili, K. N. Yang, and M. B. Maple, Phys. Rev. B 37:2368 (1988).
7. B. D. Dunlap, M. Slaski, D. G. Hinks, L. Soderholm, M. Beno, K. Zhang, C. Segre, G. W. Crabtree, W. K. Kwok, S. K. Malik, I. K. Schuller, J. D. Jorgensen, and Z. Sungaila, J. Mag. Mag. Matl. 68:L139 (1987).
8. T. Kobayashi, K. Amaya, T. Kohara, K. Ueda, Y. Kohori, Y. Oda, and K. Asayama, J. Phys. Soc. Japan 56:3805 (1987).
9. J. A. Hodges, P. Imbert, and G. Jéhanno, Solid State Commun. 64:1209 (1987).
10. See reference 6 for other work cited therein.
11. D. McK. Paul, H. A. Mook, A. W. Hewat, B. C. Sales, L. A. Boatner, J. R. Thompson, and M. Mostoller, Phys. Rev. B 37:2341 (1988).
12. J. W. Lynn, W.-H. Li, Q. Li, H. C. Ku, H. D. Yang, and R. N. Shelton, Phys. Rev. B 36:2374 (1987).
13. A. I. Goldman, B. X. Yang, J. Tranquada, J. E. Crow, and C.-S. Jee, Phys. Rev. B 36:7234 (1987).
14. P. Fischer, K. Kakurai, M. Steiner, K. N. Clausen, B. Lebech, F. Hulliger, H. R. Ott, P. Brüesch, and P. Unternährer, Physica C 152:145 (1888).
15. K. N. Yang, J. M. Ferreira, B. W. Lee, M. B. Maple, W.-H. Li, J. W. Lynn, and R. W. Erwin, Phys. Rev. B (to be published).
16. J. D. Jorgensen, M. A. Beno, D. G. Hinks, L. Soderholm, K. J. Volin, R. L. Hitterman, J. D. Grace, I. K. Schuller, C. U. Segre, K. Zhang, and M. S. Kleefisch, Phys. Rev. B 36:3608 (1987).
17. J. R. Thompson, B. C. Sales, Y. C. Kim, S. T. Sekula, L. A. Boatner, J. Brynestad, and D. K. Christen, Phys. Rev. B 37:9395 (1988).
18. J. A. Hodges, P. Imbert, J. B. Marimon Da Cunha, J. Hammann, E. Vincent, and J. P. Sanchez, Physica C 156:143 (1988).
19. I. Oguro, T. Tamegai, and Y. Iye, Physica B 148: 456 (1987).
20. S. Waki, Y. Yamaguchi, K. Oka, H. Unoki, and K. Mitsugi, Jpn. J. Appl. Phys. 26:2097 (1987).

21. C. Meyer, H.-J. Bornemann, H. Schmidt, R. Ashrens, D. Ewert, B. Renker, and G. Czjzek, J. Phys. F 17:L345 (1987).
22. B. D. Dunlap, M. Slaski, Z. Sungaila, D. G. Hinks, K. Zhang, C. Segre, S. K. Malik, and E. E. Alp, Phys. Rev. B 37:592 (1987).
23. M. B. Maple, J. M. Ferreira, R. R. Hake, B. W. Lee, J. J. Neumeier, C. L. Seaman, K. N. Yang, and H. Zhou, J. Less-Common Metals 149:405 (1989).
24. Y. Nakazawa, M. Ishikawa, and T. Takabatake, Physica B 148:404 (1987).
25. K. N. Yang, Y. Dalichaouch, J. M. Ferreira, B. W. Lee, J. J. Neumeier, M. S. Torikachvili, H. Zhou, M. B. Maple, and R. R. Hake, Solid State Commun. 63:515 (1987).
26. B. W. Lee, S. Ghamaty, and M. B. Maple (in preparation).
27. S. Simizu, G. H. Bellesis, J. Lukin, S. A. Friedberg, H. S. Lessure, S. M. Fine, and M. Greenblatt, Phys. Rev. B 39:9099 (1989).
28. A. J. Freeman, J. Yu, S. Massidda, and D. D. Koelling, Physica B 148:212 (1987).
29. J. T. Markert, T. W. Noh, S. E. Russek, and R. M. Cotts, Solid State Comm. 63:847 (1987).
30. E. E. Alp, L. Soderholm, G. K. Shenoy, D. G. Hinks, D. W. Capone II, K. Zhang, and B. D. Dunlap, Phys. Rev. B 36:8910 (1987).
31. S. H. Liu, Phys. Rev. B 37:7470 (1988).
32. A. P. Ramirez, L. F. Schneemeyer, and J. V. Waszczak, Phys. Rev. B 36:7145 (1987).

SPIN AND HOLE DYNAMICS IN DOPED ANISOTROPIC

HEISENBERG ANTIFERROMAGNETS

Miguel Lagos

Departamento de Física
Facultad de Ciencias, Universidad de Chile
Casilla 653, Santiago, Chile

Though the nature and physical origin of the antiferromagnetism observed in the layered perovskites remain unsettled questions, it is widely accepted that most of the action occurs in the Cu-O layers characteristic of all the compounds having a superconducting phase. The observation of long ranged two dimensional antiferromagnetic correlations far beyond the Néel temperature T_N support this view.[1] This way, the Cu-O layers are of special interest because likely provide a good realization of a low-dimensional spin-1/2 Heisenberg antiferromagnet. Although the ability of the two dimensional Heisenberg model to fully explain the magnetic properties of the layered metal oxides has not been established,[2,3] its study has became one of the most active fields of research in solid state physics.[2-7]

All the compounds of interest exhibit compositional unstabilities and admit large deviations from stoichiometry. The oxigen content at equilibrium seems to be a function of temperature, pressure and doping. The variable oxigen content is certainly associated to valence unstabilities in at least one of the stable atomic species present in the unit cell. Whether the charge fluctuations do occur in the Cu-O layers or in between them is not clear yet. However, it seems well established by experiments on the basic compound La_2CuO_4 with different dopages and oxigen vacancies that the relevant variable for most physical properties is the hole concentration.[8] The role of the doping by divalent cations, which substitute the trivalent rare earth, thus seems to be to compensate the excess electrons associated to oxigen defect, making holes to predominate. On the other hand, while the Cu sublattice is entirely responsible for the antiferromagnetic order, holes seem to locate in the oxigens.

It is apparent from the preceding discussion that the study of the dynamics of doped Heisenberg antiferromagnets is of current interest. The recent literature shows a number of works giving to the magnetic interactions a leading role in the phenomenology of the layered superconductors.[9] Keeping in mind the oxide superconductors, but with less ambitious purposes, we focus here on the problem of the anisotropic Heisenberg antiferromagnet, with anisotropy towards the Ising sector, interacting with a dilute fermion gas. There is no evidence that the magnetic interactions in the Cu-O planes of the layered metal oxides has such anisotropy, and our aim is simply to give an insight on the properties of a model which may be of interest for the understanding of the physics of the oxide superconductors.

The exact solution of the model alluded above seems at present hopeless. Instead, we try an approximate formalism[10,11] which, though sacrificing some numerical precision, has proven to be practical and quite reliable.[12,13] The analytic approach to the anisotropic Heisenberg model with antiferromagnetic coupling employed below has been developed very recently. It has the advantage of combining mathematical simplicity, even when working with two- and three-dimensional lattices of different symmetries,[11] with high accuracy when keeping the anisotropy parameter within the domain of validity of the theory.[12,13] Though holding in principle for the regime of high antiferromagnetic correlations, the formalism turns out to yield excellent results over an unexpectedly wide range of the anisotropy parameter α ($\alpha = 0$ and $\alpha = 1$ correspond to the Ising and isotropic Heisenberg models, respectively). Comparisons with the results of elaborate computer calculations reveal that the ground state of the asymptotic analytic theory is accurate to 0.5% for $0 \leq \alpha \leq 0.5$ in one and two dimensions.[12,13] Moreover, reasonable numerical precision in most calculations is achieved for $0.5 \leq \alpha \leq 1$ and two dimensions.[13] The mathematical simplicity of the method makes it suitable for dealing with the Heisenberg model with additional complications.[14]

We assume a lattice of spins with no frustration of the antiferromagnetic order. Thus, it can be separated into two crystallographically equivalent sublattices. One of them (associated $e.g.$ to spin up) is described by a set of vectors $\vec{R}$. The other one (spin down) is characterized by the vectors $\vec{R} + \vec{\delta}$, where vectors $\vec{\delta}$ connect each lattice site with its z nearest neighbors. A translation in $\vec{\delta}$ always implies a change of sublattice, but $\vec{R} + \vec{\delta} + \vec{\delta}'$ and $\vec{R}$ are in the same sublattice for any $\vec{\delta}$ and $\vec{\delta}'$.

The model Hamiltonian then reads

$$H = H_{\mathrm{s}} + H_{\mathrm{h}} + H_{\mathrm{int}} \ , \tag{1}$$

where H_{s} is the anisotropic Heisenberg Hamiltonian

$$H_{\mathrm{s}} = J \sum_{\vec{R}\vec{\delta}} \left[s_z(\vec{R} + \vec{\delta}) s_z(\vec{R}) + \frac{\alpha}{2} \left(s_+(\vec{R} + \vec{\delta}) s_-(\vec{R}) + s_+(\vec{R}) s_-(\vec{R} + \vec{\delta}) \right) \right] \ , \tag{2}$$

s_x, s_y, and s_z denote spin components associated to the lattice site stated in their arguments, $s_\pm = s_x \pm i s_y$, J is a positive constant and α is the anisotropy parameter. The number of sites in the spin lattice is denoted N.

The term

$$H_{\mathrm{h}} = \sum_{\vec{k}\sigma} \epsilon_k \, a_{\vec{k}\sigma}^\dagger a_{\vec{k}\sigma} \tag{3}$$

describes the dynamics of the η holes moving in a different lattice, but having same symmetry than the spin lattice. The operators $a_{\vec{k}\sigma}^\dagger$ destroy Bloch one-electron states which, due to the spin antialignment, have twice the crystal periodicity. Vector $\vec{k}$ runs over the first Brillouin zone of one of the sublattices. Therefore, the one-particle basis set incorporates the interaction of the holes with a rigid configuration of antialigned spins. The energy exchange between holes and spins are governed by the interaction term H_{int}. One of the main aspects of this work is the discussion of the general form of H_{int} and its physical implications.

It has been shown in previous papers that the manifold of stationary states of zero total spin of the antiferromagnetic Heisenberg model, with $0 \leq \alpha \leq 0.5$, can be fully described by the set of dynamical variables[10−13]

$$\phi_{\vec{\delta}}^\dagger(\vec{k}) = \sqrt{\frac{2}{N}} \sum_{\vec{R}} e^{i\vec{k}\cdot\vec{R}} \, s_+ (\vec{R} + \vec{\delta}) s_- (\vec{R}) + \frac{\alpha}{2(z-1)} \sqrt{\frac{N}{2}} \, \delta_{\vec{k},\vec{0}} , \qquad (4)$$

or Hermitian linear combinations of them. The dynamical coupling between the antiferromagnetic order and the hole extended states can thus be expressed by a potential energy of interaction having the general form

$$H_{\text{int}} = V_\sigma \left(\vec{r}, \phi_{\vec{\delta}}(\vec{k}), \phi_{\vec{\delta}}^\dagger(\vec{k}) \right) , \qquad (5)$$

where $\vec{r}$ denotes the position of a hole. It has also been shown that for $0 \leq \alpha \leq 0.5$ the Heisenberg Hamiltonian H_s can be written as[10,11,13]

$$H_s = J \sum_{\vec{k}\vec{\delta}} \phi_{\vec{\delta}}^\dagger(\vec{k}) \, \phi_{\vec{\delta}}(\vec{k}) + E_0(\alpha) , \qquad (6)$$

where $E_0(\alpha)$ is the ground state energy of H_s

In fact the operators (4) are more general than those defined previously. The generalized operators (4) reduce to the old ones for $\vec{k} = 0$. However, since only the operators $\phi_{\vec{\delta}}^\dagger(0)$ and $\phi_{\vec{\delta}}(0)$ contribute to the ground state of the Heisenberg part of H, all the results and conclusions of the series of papers in which the analytical method used here was developed[10,11] and tested with numerical calculations[12,13] do apply, no matter the generalization introduced by Eq. (4). Though not affecting the ground state of H_s, the operators (4) with $\vec{k} \neq \vec{0}$ allow to account for the magnetic excitations in a more complete way.

Going now to a weak spin-hole coupling scheme one can expand the right hand side of Eq. (5) and retain up to first order terms and write

$$H_{\text{int}} = U \sqrt{\frac{2}{N}} \sum_{\vec{k}\vec{q}\vec{\delta}\sigma} a_{\vec{k}+\vec{q}\sigma}^\dagger a_{\vec{k}\sigma} \left(\phi_{\vec{\delta}}(\vec{q}) + \phi_{\vec{\delta}}^\dagger(-\vec{q}) \right) . \qquad (7)$$

Since for small enough α (i.e. $0 \leq \alpha \leq 0.5$) the ϕ-operators obey Bose commutation relations and are asymptotically independent,[10,11] on substituting Eqs. (6) and (7) the total Hamiltonian (1) becomes isomorphic with Frölich's one.

The electron-phonon interaction and the phenomenology derived from it has been extensively studied for decades, hence a lot of physics relative to the charge transport in antiferromagnetic crystals can be drawn using the analogy between the general form of the transformed Hamiltonian of our model and the Frölich Hamiltonian. For example, thinking by analogy one realizes that self-trapping of the band holes will occur when their coupling U with the antiferromagnetically correlated electrons be large enough. A mechanism of magnetically driven charge localization is thus predicted. Hole-hole attraction of magnetic origin leading to a superconducting phase may also be expected for weaker coupling.

A useful alternate expression for H can be obtained by replacing in Eq. (7) the definition (4) of the ϕ-operators and the Fourier expansion

$$\rho_\sigma(\vec{R}) = \frac{2}{N} \sum_{\vec{q}} e^{-i\vec{q}\cdot\vec{R}} \sum_{\vec{k}} a^\dagger_{\vec{k}+\vec{q}\sigma} a_{\vec{k}\sigma} \tag{8}$$

of the hole density operator $\rho_\sigma(\vec{r}) = \Psi^\dagger(\vec{r})\Psi(\vec{r})$. One is left with

$$H_{\text{int}} = U \sum_{\vec{R}\vec{\delta}s} \rho_\sigma(\vec{R})\left[s_+(\vec{R}+\vec{\delta})s_-(\vec{R}) + s_+(\vec{R})s_-(\vec{R}+\vec{\delta})\right]$$

$$+ \frac{z}{z-1}\alpha U \eta. \tag{9}$$

If the dynamics of holes is much faster than that of spins one can substitute $\rho_\sigma(\vec{R})$ in Eq. (4) by its mean value $\langle\rho_\sigma(\vec{R})\rangle = \eta/N$. The interaction term H_{int} thus takes the form of an XY Heisenberg Hamiltonian. This way, assuming mean field for the hole variables, the total Hamiltonian H recovers the anisotropic Heisenberg form, but with enhanced XY term. The anisotropy parameter changes according to

$$\alpha \to \alpha'(\eta) = \alpha + \frac{2U}{NJ}\eta. \tag{10}$$

Therefore, as long as holes move faster than spin configurations, vacancies have the effect of shifting the anisotropy towards the XY sector. This way the antiferromagnetic order is reduced and, for large enough η, the system goes through a paramagnetic regime. This conclusion is independent of the approximate equation (6), but involves a mean field approximation whose accuracy is not guaranteed in general. In any event, one can expect from it to provide an insight on the gross features of the phenomenology of the system.

The Hamiltonian of the mean field version of the model reads

$$H_{\mathrm{MF}} = H_s \left(\alpha + \frac{2U}{NJ}\, \eta \right) + \frac{z}{z-1}\alpha U \eta + \mu\eta \, , \tag{11}$$

where $H_s(\alpha')$ denotes the Heisenberg Hamiltonian with anisotropy α', and μ is the energy cost for creating a hole. The Néel temperature then varies with the hole concentration as

$$T_N(\eta) = \theta_N \left(\alpha + \frac{2U}{NJ}\, \eta \right) \, , \tag{12}$$

where $\theta_N(\alpha')$ is the Néel temperature of the antiferromagnetic Heisenberg model with anisotropy parameter α'.

The ground state energy of $H_s(\alpha')$ diminishes with α'^2 for $0 \leq \alpha' < 1$ and linearly for $\alpha' > 1$. Therefore, if $U, \mu > 0$ it is not difficult to realize that the ground state energy of H_{MF} has a minimum for some value $\eta = \eta_0$ when μ is within a certain range. This minimum determines the equilibrium concentration of holes at $T = 0$.

This work has received financial support from **FONDECYT** (Chile).

REFERENCES

1. G. Shirane, Y. Endoh, R. J. Birgeneau, M. A. Kastner, Y. Hidaka, M. Hoda, M. Suzuki, and T. Murakami, *Phys. Rev. Lett.* **59**, 1613 (1987).
2. S. Chakraverty, B. I. Halperin, and D. R. Nelson, *Phys. Rev. Lett.* **60**, 1057 (1988).
3. D. R. Grempel, *Phys. Rev. Lett.* **61**, 1041 (1988).
4. S. Liang, B. Doucot, and P. W. Anderson, *Phys. Rev. Lett.* **61**, 365 (1988), and references therein.
5. A. Auerbach and D. P. Arovas, *Phys. Rev. Lett.* **61**, 617 (1988).
6. Y. Okabe, M. Kikuchi, and A. D. S. Nagi, *Phys. Rev. Lett.* **61**, 2971 (1988).
7. D. C. Mattis and C. Y. Pan, *Phys. Rev. Lett.* **61**, 463 (1988).
8. J. B. Torrance Y. Tokura, A. I. Nazzal, A. Bezinge, T. C. Huang, and S. S. P. Parkin, *Phys. Rev. Lett.* **61**, 1127 (1988).
9. J. R. Schrieffer, X. -G. Wen, and S. -C. Zhang, *Phys. Rev. Lett.* **60**, 944 (1988); B. I. Shraiman and E. D. Siggia, *Phys. Rev. Lett.* **61**, 467 (1988).
10. M. Lagos and G. G. Cabrera, *Phys. Rev. B* **38**, 659 (1988).
11. G. G. Cabrera, M. Lagos, and M. Kiwi, *Solid State Commun.* **68**, 743 (1988).
12. M. Lagos, M. Kiwi, E. R. Gagliano, and G. G. Cabrera, *Solid State Commun.* **67**, 225 (1988).
13. M. Lagos M. Kiwi, E. R. Gagliano, and G. G. Cabrera, *Solid State Commun.* **70**, 431 (1989).
14. D. Gottlieb and M. Lagos, *Phys. Rev. B* **39**, 2960 (1989).

SPIN-HOLE MODEL WITH MAGNETIC VORTEX-ANTIVORTEX

PAIRING MECHANISM FOR DOPED La_2CuO_4

A. Robledo

Instituto de Física,
Universidad Nacional Autónoma de México
Apartado Postal 20-364, 01000 México, D.F.

C. Varea

Facultad de Química
Universidad Nacional Autónoma de México
04510 México, D.F.

INTRODUCTION

The introduction of charge carriers in La_2CuO_4 can be achieved either by raising oxygen stoichiometry or by cation substitution, typically via the replacement of La^{+3} by Sr^{+2} or Ba^{+2} .[1] The presence of charge carriers, which appear as electron holes on the CuO_2 planes, has a profound effect on the properties of the system. Increasing their numbers beyond a low-lying threshold ($x \approx 0.06$ in $La_{2-x}Sr_xCuO_4$) transforms the initial insulating (I) (and antiferromagnetic (AF) at low temperature) state into a metallic (M) (superconductive (SC) at low temperature) state, and this exhibits an unusual, and not yet fully characterized, magnetic behavior.[2] Doping induces, too, structural changes, the most aparent of which is a transformation from an orthorhombic (O) to a tetragonal (T) structure observable at temperatures as high as 530 K when close to the undoped limit. [1,2] There are indications of other, still incompletely resolved transformations,[3] closer to, or within, the SC region of the phase diagram. The interplay between superconductivity, magnetism and structure in doped La_2CuO_4 is the focus of continuing investigations.

Here we describe features of a spin-hole 2D Heisenberg model[4] with a weak anisotropy that is transformed with charge carrier concentration from Ising to XY type, and show that its properties are consistent with those of doped La_2CuO_4 . When holes are localized (at O sites on the CuO_2 planes) the Cu spin system is represented by competing AF and F *fixed* couplings. [5] But when holes are mobile the Cu spin system becomes one of *averaged* and *uniform* couplings, the magnitude and anisotropy of which are determined by hole concentration. The key element in the predicted phase diagram is a critical dopant fraction x_0 at which the Cu-Cu spin interactions switch from AF to F *global* character, from (canted) Ising to (canted) XY anisotropy, and from electrical insulator to conductor. Conduction charges are paired via the net attraction of the Kosterlitz-Thouless (KT) vortex excitations to which they become associated. We have estimated the KT transition temperature and the size of vortex pairs dressed with charges from the measured values of the spin coupling constants in the undoped limit. These are in good agreement with the observed SC temperature and coherence length in the La_2CuO_4-based system.

HEISENBERG HAMILTONIAN WITH ANTISYMMETRIC EXCHANGE

The orthorhombicity of La_2CuO_4 below 530 K is due to the buckling of the CuO_2 planes that results from rotation of the CuO_6 octahedra.[1,2] For this reason nearest-neighbor Cu ions on adjacent planes are not all equidistant. The 3D AF ordering observed in La_2CuO_4 (in contrast with the 2D AF long range order (LRO) of the isostructural, but tetragonal, K_2NiF_4) derives from the disbalance in the weak interlayer coupling introduced by the orthorhombic symmetry. Even though this interlayer coupling is small ($J_\perp \sim 0.003$ meV), [2] the undoped system exhibits a large value for T_{3D}^{AF} ($\sim$ 300 K) because of the unusually large in-plane correlation length ξ_{2D}. [2] The value of ξ_{2D} is determined by the strong isotropic component of the in-plane couplings J (J $\sim$116 meV),[2] but the transition to 3D LRO appears to be driven by $J_\perp$.[2] As we quote below, $J_\perp$ is of the same order of the Ising anisotropy of the intraplanar couplings. As it is known, [6] 2D Heisenberg systems are extremely sensitive to anisotropy , and vanishingly small amounts are sufficient to lift the transition temperature to finite values for both uniaxial and planar anisotropy. [6]

The detailed magnetic structure of pure La_2CuO_4 is now better understood after the discovery of a field-induced transition below T_{3D}^{AF} to a state with weak F LRO.[1,2] This behavior is due to the canting of spins that occurs when there are antisymmetric terms in the exchange Hamiltonian. [1,2] The rotated CuO_6 octahedra in the orthorhombic phase of La_2CuO_4 is responsible, not only for a nonvanishing nearest-neighbor contribution to $J_\perp$, but also for the antisymmetric superexchange terms in the intraplanar Cu-Cu spin interaction, which, as we quote below, are the dominant corrections to the isotropic coupling, and play an important role in our model phase diagram.

In the undoped system the antisymmetric superexchange interactions arising from the Cu-O-Cu complexes are represented by the spin Hamiltonian [2]

$$\mathcal{H} = \sum_{\langle i,j \rangle} S_i \cdot J \cdot S_{j'} \tag{1a}$$

where $\langle i,j \rangle$ indicates summation over nearest neighbor pairs only, where

$$J = \begin{pmatrix} J_{aa} & 0 & 0 \\ 0 & J_{bb} & J_{bc} \\ 0 & -J_{bc} & J_{cc} \end{pmatrix}, \tag{1b}$$

and where S_i and S_j represent Heisenberg spins on the Cu sites, and a, b and c indicate the La_2CuO_4 unit cell directions, with b normal to the basal plane.

We consider the planar lattice of Cu sites subdivided into two interwoven sublattices of sites of types P and Q. Now we employ two Cartesian frames of reference (a'b'c') and (a"b"c") obtained from that defining the crystal unit cell (abc) through rotation by an angle ϑ in the b direction away from c when site i is of type P and by an angle $-\vartheta$ when site i is of type Q, respectively. Eq. 1b can be written in the form

$$J \cdot = \begin{pmatrix} J_{a'a''} & 0 & 0 \\ 0 & J_{b'b''} & 0 \\ 0 & 0 & J_{c'c''} \end{pmatrix}, \tag{2a}$$

where

$$J_{a'a''} = J_{aa}, \tag{2b}$$

$$J_{b'b''} = \frac{1}{2}(J_{bb} - J_{cc}) + \frac{1}{2}[\,\text{sgn}\,(J_{bb}+J_{cc})]\,[\,(J_{bb}+J_{cc})^2 + (2J_{bc})^2\,]^{1/2} \tag{2c}$$

$$J_{c'c''} = \frac{1}{2}(J_{cc} - J_{bb}) + \frac{1}{2}[\,\text{sgn}\,(J_{bb}+J_{cc})]\,[\,(J_{bb}+J_{cc})^2 + (2J_{bc})^2\,]^{1/2} \tag{2d}$$

when the angle ϑ (in radians) is given by

$$\vartheta = \frac{1}{2}\,\text{arctg}\,[\,2J_{bc}\,(J_{bb}+J_{cc})^{-1}\,]. \tag{2e}$$

The relative values of $J_{a'a''}$, $J_{b'b''}$ and $J_{c'c''}$ in Eqs. 2 determine the type of anisotropy imprinted on the 2D Heisenberg magnet. This has been quantified[2] for undoped La_2CuO_4, the values deduced from experiment are $J \sim 116$ meV, $J_{bc} \sim 0.55$ meV and $J_{bb} - J_{cc} \sim 0.004$ meV, with $J_{aa} - J_{cc} \ll J_{bb} - J_{cc}$, from which one obtains weak Ising anisotropy, $K^{AF} = J_{c'c''} - J \sim 0.003$ meV (where $J = 1/3(J_{a'a''} + J_{b'b''} + J_{c'c''})$), along the c' axis with a small canting angle $\vartheta \sim 0.003$. It is to be noted that the anisotropy K^{AF} is approximately equal to the interplanar coupling $J_\perp$ (therefore the 3D nature of the onset of AF LRO) and these turn out to be very small when compared with the isotropic J. However, the ratio $J_{bc}/J \sim 0.005$ is of the order of magnitude of the degree of anisotropies found in the isostructural 2D magnets K_2NiF_4 and K_2CuF_4.[4]

The Hamiltonian in Eq. (1) applies only to the undoped system. As mentioned, the introduction of a hole localized at an oxygen site changes the local Cu-Cu spin interaction from AF to F character. We use below superindices AF and F to distinguish between the two sets of local Cu-Cu couplings. The presence of a hole not only changes the sign of the isotropic term J but also perhaps the relative values of $J_{a'a''}$, $J_{b'b''}$ and $J_{c'c''}$. In case that the local F coupling presents XY anisotropy, addition of holes introduces not only competing AF and F interactions, but also competing anisotropies. Since the experimental observations in $La_{2-x}Sr_x CuO_4$ indicate that AF LRO is very quickly destroyed with doping, the suppression of AF LRO may not only be due to frustration, but also to an improved approximation to the isotropic Heisenberg limit as the result of an over-all cancellation of anisotropy. (Hole mobility is another factor that contributes to a fast decrement of the Néel temperature because it may enhance the effective number of F couplings beyond the nominal value of x, and, furthermore, produce a more uniform spatial variation of anisotropy). As we shall see below, the main contribution to *global* XY anisotropy at high dopant fraction originates from the antisymmetric term $J_{bc}^{AF} \sim J_{bc}^F$. Nevertheless, the competition between the small opposing local anisotropies may contribute to the fast disappearance of the Néel state with x.

In modelling a spin-hole system compatible with the Hamiltonian for the undoped system in Eq. (1) we adopt two simple limiting situations. In the first one, that considers holes to be firmly trapped (the insulating lowly-doped limit) we take up an anisotropic Heisenberg model with a random distribution (aside from the correlations that arise from the repulsion between holes) of *fixed* AF and F couplings. The thermodynamic behavior in this case is reminiscent of that of random systems with competing interactions.[7] In the second limit, appropriate when holes are extremely mobile (the highly-doped conducting phase) the system is represented by *uniform* average couplings with values that depend on the dopant fraction. Here the free energy for the magnetic structure of the Cu spins is much simpler than that obtained for fixed holes, and corresponds to that of conventional spatially-uniform couplings. In this case holes are assumed to become a charge carrier fluid whose spins only affect the magnetic structure of the system via the introduction of a homogeneous ferromagnetic component to the Cu couplings. Apart from the background AF couplings this picture is similar to that of itinerant magnetism in transition metal elements. These two limiting situations are briefly discussed below.

ISING ANISOTROPY REGIME

Recently,[8] we have obtained the effect of competition between AF and F couplings on the correlation length (for a simple mean-field spin-hole model with classical and purely Ising spins), i.e. an expression for ξ as a function of hole concentration n_h. In the low dopant concentration interval $0 < n_h < (1 + (J^F /|J^{AF}|)^2)^{-1}$ the correlation length is of AF-type[8] and is given by

$$\xi^{AF} = - \left\{ \ln \left[\left(\frac{R}{2S} - 1 \right) - \sqrt{ \left(\frac{R}{2S} - 1 \right)^2 - 1 } \right] \right\}^{-1} \tag{3a}$$

where

$$R = (kT)^2 - 8[(J^{AF})^2 n_b + (J^F)^2 n_h] \quad \text{and} \quad S = 2[(J^{AF})^2 n_b - (J^F)^2 n_h], \tag{3b}$$

where $n_b = 1 - n_h$. Recalling the known [6] crossover properties of the anisotropic Heisenberg model with classical spins, we can make use of the above expression for ξ in the thermodynamic description of the CuO_2 layers. We note that for small anisotropy in mean-field approximation we can write[4]

$$kT_{2D}^{AF} = (\xi_H /a)^2 \, K^{AF} (1 - n_h)^{1/2}, \tag{4}$$

where T_{2D}^{AF} is the Néel temperature of the uncoupled CuO_2 layers with small Ising anisotropy K^{AF} and ξ_H is the correlation length of the isotropic 2D Heisenberg system. Since the correlation length decreases because of the frustration that proliferation of F couplings introduces, it is reasonable to assume, within the level of our simple and classical mean-field description, that ξ_H decreases as ξ^{AF} does. We can make use of Eqs. 3 with appropriate choices of parameters, and for this purpose we choose, in the abscence of more precise information, $J^F/|J^{AF}| = 1.1$, and we fix the temperature when determining the variation of ξ_H with hole concentration such that T_{2D}^{AF} is 300 K for the undoped system. In any case, $\xi_H = 0$ when $n_h = (1 + (J^F /|J^{AF}|)^2)^{-1}$.

We prescribe now a relation between n_h and x. In our treatment the percolation threshold for AF LRO is (unrealistically) given by $n_h^{AF} = 1$, the effect of competing anisotropies and the hole localization length have not been considered, hence n_h and x are expected to differ. To make some comparison with experiment we incorporate these features phenomenologically into the model. We replace the hole density n_h by an effective quantity given by $n_h = (1/2)[1 - \exp(-t^2x)]$, with $n_b = 1 - n_h$. This expression, in agreement with experimental observations,[1] allows n_h to increase linearly for small x, as $n_h = t^2x/2$, but it saturates to a value $n_h^{max} \sim 1/2$ for larger x. Contrary to the cases of the Y-Ba-Cu-O and Bi-Y-Ca-Cu-O systems, the I-M transition in the La-Sr-Cu-O system occurs for a very small value of x, $x_o \sim 0.06$, and in our model this requires a pronounced initial increment of n_h with x. When x is small $n_h \sim t^2x/2$, and therefore we note that the quantity $t/\sqrt{2}$ can be interpreted as a localization length for the holes (i.e. x is dopant occupation per unit planar lattice cell and $t^2/2$ the number of such cells occupied per hole). This localization length has been estimated [9] to be of the order of 2-3 lattice constants. With the choice $t = 6$ ξ_H vanishes at $x_o = 0.06$. The value given to the Ising anisotropy K^{AF} in Eq. 4 is 0.003meV. In Fig. 1 we show our results for T_{2D}^{AF} .

Since $J_\perp$ is small when compared to the isotropic intraplanar couplings, the variation of the 3D Néel temperature with x can be estimated in the same way as we did above for T_{2D}^{AF}. Since the dominant contribution to the intraplanar correlation length is ξ_H we have $T_{3D}^{AF} \approx (\xi_H/a)^2 J_\perp$. $J_\perp$ is approximately equal to the anisotropy K^{AF} and the 2D and 3D transition temperatures turn out to be indistinguishable in the scale of Fig. 1. Therefore we obtain a reasonable description for the onset of AF LRO at low T with the Néel temperature decreasing rapidly with increasing doping. We note in Fig. 1 that, after its fast decrement, the Néel temperature shows a "foot" before vanishing at $x_o \sim 0.06$. It has been suggested [2] that this region corresponds to one of spin glass behavior in doped La_2CuO_4. Like in many spin models with random positional distribution of AF and F interactions, frustration leads to spin glass behavior at low temperature (provided $T_{2D}^{AF} \approx 0$), but the simple methods we employed here do not provide information on this respect.

The mean-field Eqs. 3 indicate how the correlation length ξ in the paramagnetic phase of the spin-hole model with fixed AF and F local couplings is transformed with doping. The effect of frustration generated by increasing numbers of F couplings on $\xi = \xi^{AF}$ is to decrease its value until it vanishes at a critical dopant fraction x_o for all temperatures (i.e. when S = 0). The small coupling between the CuO_2 layers implies that a directionally-averaged ξ does not actually vanish at x_o but may become very small in its neighborhood. In any case, it is important to notice that the intralayer spin fluctuations are of AF type when $x < x_o$ and of F type when $x > x_o$. See Fig. 1.

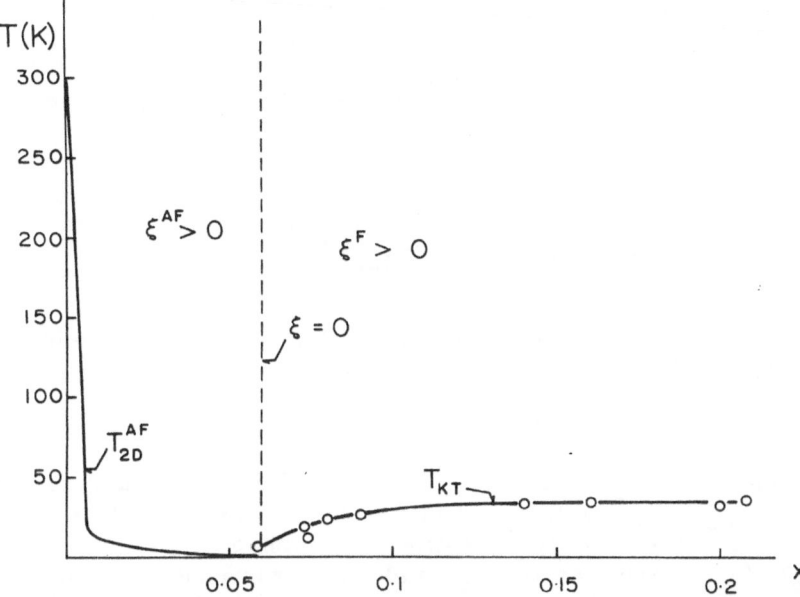

Fig. 1 Calculated phase diagram for an uncoupled spin-hole model layer. At $x = x_0$ the system switches from AF to F character and the correlation length ξ vanishes. Since an F background favours charge conductivity whereas an AF one does the opposite we identify the line $x = x_0$ as the I-M transition line. T_{2D}^{AF} and T_{KT} correspond to the Néel and KT transition temperatures, respectively. The circles are experimental data taken from Ref. 1.

XY-ANISOTROPY REGIME

We identify the loci at which the correlation length ξ of an uncoupled model CuO_2 layer vanishes, and a change from AF to F type in the Cu spin-spin correlation occurs, as the model's representation of the insulator-to-metal transition in the copper oxides. There, the strongly coupled spins behave, on average, as independent spins (since $\xi = 0$) and the overturning of spins that the AF environment imposes to in-plane hole translation disappears. Beyond x_0, the prevailing F-like correlations enhance hole conductivity. Furthermore, large and sudden changes in hole conductance can have a strong effect in the magnetic properties of the system. Once the holes acquire a high mobility our picture of randomly distributed, and *permanent* for practical purposes, AF and F spin couplings gives way to one of *averaged* and *uniform* couplings $<J>$. The Cu-spin Hamiltonian would be now given by

$$\mathcal{H} = \sum_{\langle i,j \rangle} S_i \cdot <J> \cdot S_j , \tag{5a}$$

where

$$<J> = \begin{pmatrix} \langle J_{aa} \rangle & 0 & 0 \\ 0 & \langle J_{bb} \rangle & \langle J_{bc} \rangle \\ 0 & -\langle J_{bc} \rangle & \langle J_{cc} \rangle \end{pmatrix} . \tag{5b}$$

A simple mean-field estimate of such averaged diagonal coupling $\langle J_{aa} \rangle$ at dopant fraction $x > x_0$ is

$$\langle J_{aa} \rangle = J_{aa}^{AF} n_b^{1/2} + J_{aa}^F n_h^{1/2} , \tag{5c}$$

and similarly for $\langle J_{bb} \rangle$ and $\langle J_{cc} \rangle$. For simplicity we assume $J_{aa}^F / |J_{aa}^{AF}| = J_{bb}^F / |J_{bb}^{AF}| = J_{cc}^F / |J_{cc}^{AF}|$, and we choose this ratio to be again, as above, 1.1. The average isotropic coupling $\langle J \rangle = 1/3(\langle J_{aa} \rangle + \langle J_{bb} \rangle + \langle J_{cc} \rangle)$ vanishes at x_0, and is ferromagnetic when $x > x_0$. However, recalling that the antisymmetric term J_{bc} arises from the rotation of the CuO_6 octahedra in the orthorhombic phase, and because the La-based system exhibits appreciable and similar orthorhombic strains both before and after the I-M transition, we expect the antisymmetric coupling $\langle J_{bc} \rangle$ to have a weak dependence on x and to remain of the order of $|J_{bc}^{AF}| \sim |J_{bc}^F|$ throughout the I-M transition. It is possible that J_{bc}^{AF} and J_{bc}^F have both the same sign.

The transformation from the frame of reference (abc) into the rotated frames (a'b'c') and (a"b"c") allows us to employ Eqs. 2 with the couplings in them replaced by the above estimates for the average uniform couplings in Eqs. 5 . We obtain

$$<J> = \begin{pmatrix} \langle J_H \rangle & 0 & 0 \\ 0 & \langle J_H \rangle & 0 \\ 0 & 0 & \langle J_H \rangle \end{pmatrix} + \begin{pmatrix} 0 & 0 & 0 \\ 0 & \langle J_{XY} \rangle & 0 \\ 0 & 0 & \langle J_{XY} \rangle \end{pmatrix} + \begin{pmatrix} 0 & 0 & 0 \\ 0 & 0 & 0 \\ 0 & 0 & \langle J_I \rangle \end{pmatrix} , \tag{6}$$

where the isotropic Heisenberg, $\langle J_H \rangle$, XY, $\langle J_{XY} \rangle$, and Ising, $\langle J_I \rangle$, components above are given by

$$\langle J_H \rangle = \langle J_{aa} \rangle , \tag{7a}$$

$$\langle J_{XY} \rangle = \frac{1}{2} \left(\langle J_{bb} \rangle - \langle J_{cc} \rangle - 2\langle J_{aa} \rangle + [(\langle J_{bb} \rangle + \langle J_{cc} \rangle)^2 + 4\langle J_{bc} \rangle^2]^{1/2} \right) \tag{7b}$$

$$\langle J_I \rangle = \langle J_{cc} \rangle - \langle J_{bb} \rangle . \tag{7c}$$

The most remarkable feature that results from this replacement is that at $x = x_o$ the spin model for the Cu-Cu spins for an uncoupled CuO_2 plane becomes a purely (but canted) 2D XY model with $\langle J_H \rangle = 0$, $\langle J_{XY} \rangle = J_{bc}$ and $\langle J_I \rangle = 0$. When $x > x_o$ the terms $\langle J_H \rangle$ and $\langle J_I \rangle$ become different from zero, and the pure XY Hamiltonian gives way to one of Heisenberg with anisotropy. Since the magnitudes of the diagonal local couplings J_{aa}^F, J_{aa}^{AF}, etc. are much larger than their differences, $\langle J_H \rangle$ grows, with positive F values, much faster than $\langle J_I \rangle$. With the values we have chosen for our parameters, $\langle J_I \rangle$ is always at least five orders of magnitude smaller than $\langle J_H \rangle$. However, $\langle J_{XY} \rangle$ is for all $x > x_o$ at least three orders of magnitude larger than $\langle J_I \rangle$. When x is still very close to x_o the anisotropy is strong but it diminishes rapidly with increasing x, however it remains always dominantly of XY type.

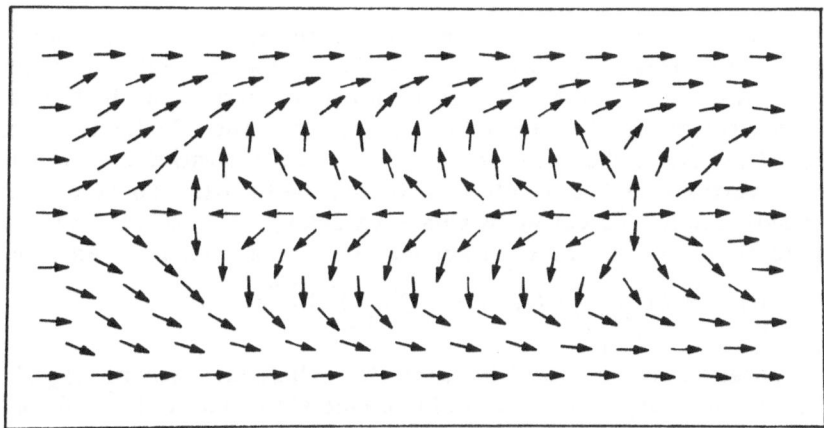

Fig. 2 A vortex-antivortex pair excitation in the F XY anisotropy regime. When charges are trapped by the AF environment at their centers they become themselves paired. The charge repulsion is overcome by the logarithmic attraction between magnetic vortices and the estimated center-to-center separation is about 20 Å.

The known properties of the 2D XY magnet,[10] or alternatively those of a 2D Heisenberg system with XY anisotropy, [6] indicate that the Cu spins of our model cannot exhibit F LRO at any temperature in the absence of an applied field when $x > x_o$. Instead, vortex-like spin fluctuations (see Fig. 2) play a central role in configuring a low temperature phase. Whereas in the high-temperature phase vortices can appear isolated from each other, they form vortex-antivortex pairs in the low temperature phase. Following the same procedure employed for

T_{2D}^{AF} above, we estimate the associated Kosterlitz-Thouless transition temperature T_{KT}. We obtain[4]

$$kT_{KT} = \pi (\xi_H /a)^2 \langle J_{XY} \rangle ,\qquad\qquad (8)$$

where the correlation length ξ_H has an exponential dependence on the isotropic Heisenberg coupling, i.e. $\xi_H = A \exp(2\pi \langle J_H \rangle B/kT)$. We choose the constants A and B to be A=0.32 and B=0.635, equal to those found to fit experimental data on La_2CuO_4 in a recent Monte Carlo simulation[11] of the 2D Heisenberg model. With the employment of the expression above for ξ_H, together with the phenomenological relation $n_h = \frac{1}{2}[1 - \exp(-tx)]$ we evaluate T_{KT} as a function of x. Even though the degree of XY anisotropy $\langle J_{XY} \rangle / \langle J_H \rangle$ drops rapidly with increasing x, the parallel growth of ξ_H makes T_{KT} increase with x. Our results are shown in Fig. 1. These are in close agreement with the measured values of the superconducting T_c in this system.[1]

As can be seen in Fig. 2, the spin configuration at the center of a vortex of the ferromagnetic XY system is locally antiferromagnetic, therefore we expect these vortices to localize charges, as in the lowly-doped antiferromagnet with Ising anisotropy. The Coulomb repulsion between the vortex charges is overcome by the logarithmic [10] attraction between vortices and one obtains vortex pairs with short center-to-center separations. To estimate this separation, we note that the energy of interaction $\varphi(r)$ of two such vortices located a distance r apart is

$$\varphi(r) = \pi \langle J_{XY} \rangle \ln (r/a) + \left(\frac{e^2}{4\pi\varepsilon\varepsilon_o} \right) \frac{e^{-K_o r}}{r}. \qquad (9)$$

The separation r_o at which $\varphi(r)$ is minimum (when $\langle J_{XY} \rangle = J_{bc}$ =0.55 meV, the transition metal oxide dielectric constant is $\varepsilon = 10$ and the Thomas-Fermi screening length is $K^{-1} = 3.6$ Å) is, according to Eq. 9, $r_o = 20.2$ Å, which is very close to the values measured for the superconducting coherence length in the La_2CuO_4 system. [1]

As described, our model suggests naturally a magnetic mechanism for the pairing of charges in doped La_2CuO_4 . (The onset of super-conductivity at T_c may coincide or be close to the Kosterlitz-Thouless T_{KT} of the magnetic system). The pairing of charges in our model has been linked to the appearence and permanence of XY anisotropy for a range of dopant fractions, and this in turn has been related to: i) the mobility of the charges in a ferromagnetic background, and, ii) the buckling of the CuO_2 planes and the ensuing orthorhombicity and canting of spins. (It is interesting to note here that recent[3] structural studies of $La_{2-x}Ba_xCuO_4$ find indications of a low-temperature tetragonal phase in a phase diagram region that envelops the SC phase.

This new T phase is interpreted as a coherent superposition of the two O variants that result from rotating the CuO_6 octahedra about the $(1\,0$-$1)$ and the (101) axis. The antisymmetric term J_{bc} would not vanish in this structure and we would obtain a magnetic description similar to that given here). Beyond O-T phase boundary the antisymmetric coupling $\langle J_{bc} \rangle$ vanishes, however, the system will retain a small XY-type anisotropy since there, like in the isostructural compounds of Ref. 7 , $\langle J_{aa} \rangle = \langle J_{cc} \rangle$. In this highly-doped regime, the repulsions between the large numbers of holes may affect the pinning of charges by the magnetic vortices, and it is possible that magnetic and charge transport properties become related now in a different way. A likely possibility is that the F XY magnet properties may appear unmasked in a metallic phase. There are some indications[12] of ferromagnetism at low temperature, ~10 K, in highly-doped $La_{2-x}Na_xCuO_4$.

We acknowledge financial support by the Programa Universitario sobre Superconductividad de Alta Temperatura de Transición (PUSCATT, UNAM).

REFERENCES

1. S.W. Cheong, J.D. Thompson, and Z. Fisk, Physica C158, 109 (1989).
2. R. J. Birgeneau and G. Shirane, in *Physical Properties of High Temperature Superconductors*, D.M. Ginsberg Ed. (World Scientific Publishing Co.) 1989.
3. J.D. Jorgensen *et al*, Phys. Rev. B 38, 11337 (1988); J.D. Axe *et al*, Phys. Rev. Lett. 62, 2751 (1989).
4. A. Robledo and C. Varea, Rev. Mex. Fis. 35, 255 (1989); in *Condensed Matter Theories* Vol. 4, J. Keller Ed. (Plenum Press) 1989; Phys. Rev. B (submitted).
5. Our model has many elements in common with earlier spin-hole models for the pairing of charges via magnetic interactions. See, for example, A. Aharony *et al*, Phys. Rev. Lett. 60, 1330 (1988)., and references therein.
6. D.R. Nelson and R.A. Pelcovits, Phys. Rev. B 16, 2191 (1977).
7. C. Dekker, A.F.M. Arts, and H.W. de Wijn, Phys. Rev. B 38, 11512 (1988); Y. Kimishima, *et al*., J. Phys. Soc. Jpn. 55, 3574 (1986); K. Katsumata, *et al*., Phys. Rev. B 25, 428 (1982); *ibid.* 37, 356 (1988).
8. A. Robledo and C. Varea, Int. J. Mod. Phys. B 1, 763 (1988).
9. M.A. Kastner *et al*, Phys. Rev. B 37, 111 (1988).
10. J.M. Kosterlitz and K.J. Thouless, J. Phys. C 6, 1181 (1973).
11. G. Gomez-Santos, J.D. Joannopoulos, and J.W. Negele, Phys. Rev. B 39, 4435 (1989).
12. M.A. Subramanian *et al*, Science 240, 495 (1988).

STUDY OF SUPERCONDUCTING ANsD PARENT PHASES BY CHEMICAL MODIFICATIONS

P. Barboux*, J.M. Tarascon, M.K. Kelly, P.F. Miceli
L.H. Greene, D.E. Aspnes, G.W. Hull and M. Giroud

Bellcore, Red Bank, NJ07701, USA

Y. Lepage and W.R. McKinnon

NRC, Ottawa, K1A0R9, Canada

ABSTRACT

The evolution of optical, magnetic and structural properties of the Y-based and Bi-based cuprate superconductors has been studied and compared with the behavior of parent phases obtained by oxygen deinsertion or by chemical substitution. We discuss the effects of the structural defects and of the oxygen disordering on the charge balance and on the physical properties.

INTRODUCTION

Since the discovery of high temperature superconductivity in cuprates most of the experimental techniques have been applied to gain a better understanding of these phases, but many questions remain to be answered about the doping mechanism in these materials as well as their electronic structure. These superconductors are built of copper-oxygen layers with a valence for copper around 2. Upon substitution such as divalent Sr for trivalent La in La_2CuO_4 or oxygen insertion like in $YBa_2Cu_3O_6$[1,2] the material becomes a delocalized mixed valence compound. The reverse can be done as well, for example by substituting trivalent rare-earth for divalent Ca in $Bi_2Sr_2Ca_1Cu_2O_{8+x}$[4], thus decreasing the amount of charge carriers[5]. In both ways, a transition between an antiferromagnetic insulator and a superconducting metal occurs[3].

One of the major difficulties in the study of these phases is

---

* On leave from CNRS Université P.etM. Curie, Paris, France

the importance of oxygen and cationic disorders on the metal-insulator transition. For example, the material $YBa_2Cu_3O_{6.4}$ may be a
superconductor or an insulator depending on the way the oxygen has
been depleted from the samples[2,6]. Similarly, because of an incommensurate structural modulation and a large amount of cationic
substitutional disorder, there is a large controversy about the
origin of the charge of Cu (and the modulation) in $Bi_2Sr_2Ca_1Cu_2O_8$
phase. Thus, it is important to understand how the charge carriers
delocalization takes place within the CuO planes, where the charge
reservoirs are located and how the charge transfer occurs between
these charge reservoirs and the conduction planes. We present here
some experimental results that address this question.

CHARGE BALANCE AND LOCALIZATION IN $YBa_2Cu_3O_{7-y}$

Oxygen can be removed from the 90K superconducting phase
$YBa_2Cu_3O_{7-y}$ down to y=1 inducing a semiconducting behavior. Around
y=0.5 (depending how oxygen has been removed) and orthorhombic-
tetragonal transition occurs. We have monitored this transition
through the optical properties of these materials in comparison
with parent phase, using a spectroscopic ellipsometer[7] over an
energy range of 1.5 eV to 5.8 eV[8]. Experiments were carried out on
polished ceramics but comparisons with thin films and single crystals have proven the technique to be reliable in the case of ceramic samples. As shown by Fig. 1a and 1b, upon removal of oxygen
two strong and relatively sharp optical features appear at 1.7 eV
and 4.1 eV[8-10].

The 4.1 eV peak is strongest at the lowest oxygen compositions. It is unaffected by the substitution of strontium for barium
and rare earths for ytrium as well. But its intensity is related
to the oxygen composition in the CuI site (oxygen removal was done
by annealing under argon at various temperatures between 450°C and
650°C and quenching[12]). Therefore, we have attributed this optical
feature to an electronic transition in the O-Cu-O complexes that
result from the oxygen vacancies in the copper CuI planes[8] (i.e.
the so-called chain sites when y=0). This assignment is confirmed
by the presence of a similar absorption at the same energy in the
Delafossite structures $MCuO_2$ (M=Cr,Y) where linear $Cu^{1+}O_2$ complexes are also present[9] (Fig. 1b). The compound K_2CuCl_3 also
possesses isolated Cl-Cu-Cl complexes and a similar optical excitation in this energy range. Note in Fig. 1a that this optical
absorption associated to Cu^{1+} is already present in the superconducting material. Since one oxygen removed from the CuI chains
yields a very unstable 3-fold coordination around the copper, some
attraction and ordering of the oxygen vacancies is to be expected,
leading to part of the material with copper having 2-fold coordination. This may immediately trap a localized Cu^{1+} within a O-Cu-O
complex even when the material has an average formal valency of
the copper above 2 (i.e. when y<0.5). Some other discussions about

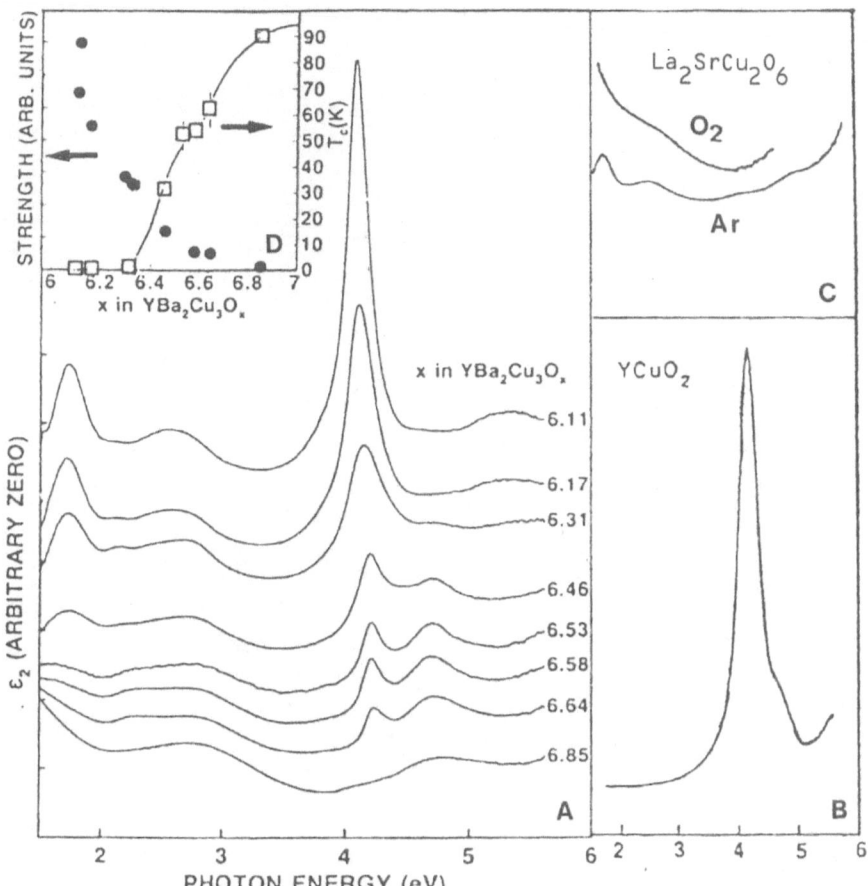

Figure 1. Measured imaginary part of the dielectric constant (a)
for different oxygen composition in $YBa_2Cu_3O_y$, (b) for
the $YCuO_2$ phase, (c) for $La_2SrCu_2O_6$ annealed under Ar
and O_2. (d) shows the intensity of the 4.1 eV peak and
the Tc associated to the different values of y. Transition temperature is defined as the transition midpoint
in the ac-susceptibility curve.

the vacancy ordering and the nature of the orthorhombic-tetragonal
transition might be found in these proceedings as well as in the
literature[11,12]. From the experimental fact that the Cu2 planes
are a common feature of all superconducting cuprates and that any
chemical substitution in these planes has a drastic effect on Tc
and on the transport properties[13] whereas it has a lesser effect
in the Cu1 planes we conclude that the transport properties are
related to the formal valency in the Cu2 planes. But, let us now

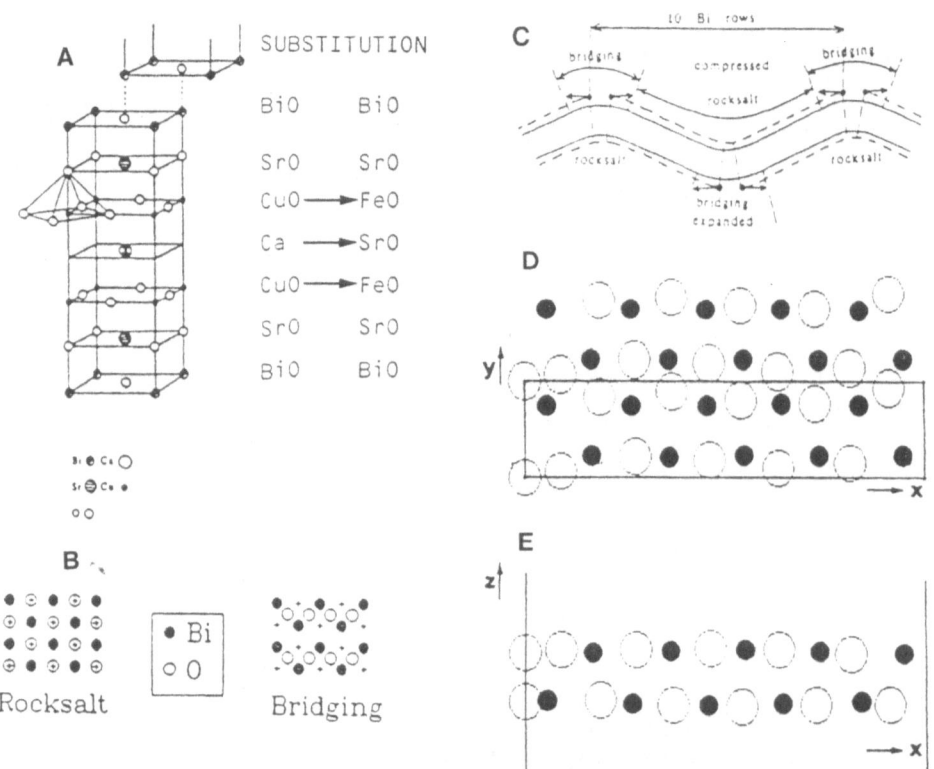

Figure 2. Structural study of the $Bi_2Sr_2CaCu_2O_{8+y}$ phase. (a) shows
how substitution of Fe for Cu occurs. (b) represents the
two competing planar arrangements for the oxygen atoms.
(c) shows the mechanism responsible for the structural
modulation after ref. 19. (d) the projection of Bi and O
onto the xy and xz planes are given for the $Bi_2Sr_3Fe_2O_{9.2}$
phase.

assume that the transport properties are related to the formal
valency in the Cu2 planes. This is consistent with the fact that
such planes are the common feature of all superconducting cuprates
and that any chemical substitution in these sites has a drastic
effect on Tc and on the transport properties[13]. Thus, the effect
of oxygen removal in the Cu1 planes which act as charge reservoirs
will depend on the configuration of oxygen vacancies. X-ray absorption measurements have shown that Cu^{1+} is formed in an amount
slightly less than y^{14}. An explanation is that for each oxygen
removed from the Cu1 planes there is a linear increase of one Cu^{1+}
and that the other charge of the oxygen is balanced by an electron
transfer to the Cu2 planes. In this case the intensity of the
4.1 eV peak should increase linearly with y which is not in agreement with the results shown in Fig. 1d. A more likely explanation
for the formation of a localized Cu^{1+} is the presence of two neigh-
bouring oxygen vacancies. Then, the amount of Cu^{1+} will equal y
for perfect ordering and y^2 if purely statistical location of
vacancies happens. Because Cu avoids 3-fold coordination clustering or ordering is favored and the amount of Cu^{1+} is nearly y but
less beacuse of oxygen disorders. This indicates that the oxygen
ordering might be as important as its stoichiometry and explains
why the observed superconducting transition temperatures depend on
the way the oxygen has been removed. An important ordering or even
clustering of the oxygen vacancies yields a large amount of Cu^{1+}
and a smaller decrease of the Cu2 site formal valency. This may
happen when oxygen has been removed at low temperature or if the
samples have been slowly cooled. In those samples superconducting
behavior is observed down to very low oxygen contents (y=0.7) in
coincidence with the observation of vacancy ordering at $y=0.5^6$.

By analogy with compounds of simillar structure we have as-
cribed the peak at 1.7 eV to a localized excitation in the copper
Cu2 planes. As a matter of fact, this absorption peak is present
in any material that possesses an extended plane of equivalent
sites of copper, that is semiconducting and that has copper coordinated by an oxygen pyramid. It is found for example in the compound $La_2SrCu_2O_6$ when semiconducting (Fig. 1c). After high oxygen
pressure annealing the peak vanishes whereas the phase becomes
conducting[9]. Therefore, this optical absorption is a localized
feature characteristic of delocalized mixed valent cuprates when
they are in their insulating form. It appears upon oxygen removal
as well as upon cationic substitution of Cu for other 3d metals.
In the case of Co that substitutes in the Cu1 site, some oxygen
disordering is also introduced. The charge transfer between the
Cu1 planes (charge reservoir) and Cu2 planes (conduction planes)
is strongly affected as found by structural studies[16] and the
decrease of the charge carriers density[18].

The charge transfer between planes affects the electronic
properties but also the magnetic properties. Another common feature

of the cupric oxides high temperature superconductors is their
antiferromagnetic behavior in the insulating form. However, there
has been some discrepancies in the results of neutron scattering
studies about the long range magnetic ordering[15]. In the first
reports only one ordering was observed with magnetic moment ordered
in the Cu2 plane sites whereas more recent works also found, at
lower temperature, a second ordering transition associated with a
magnetic moment both in the Cu2 and Cu1 sites. This is surprising
since for low oxygen content the Cu1 site is expected contain
monovalent diamagnetic Cu.

P.F. Miceli et al.[16,17] have found that the substitution in
the Cu1 site of Cu by Co increases the antiferromagnetic coupling
as well a permanent magnetic moment is stabilized on the Cu1 site.
This may explain why magnetic ordering is observed in some samples
of $YBa_2Cu_3O_{7-y}$ since it can again be related to some oxygen disorder and localization of magnetic Cu^{2+} in the Cu1 site that pins
down some localized magnetic moment.

As a conclusion, oxygen disordering may lead to different
charge balance and different physical properties associated with
a same stoichiometry.

OXYGEN STOICHIOMETRY IN $Bi_2Sr_2Ca_1Cu_2O_{8+x}$

Of all the high Tc copper oxides the Bi-based superconductors
present the most complex structure. This arises from the incommensurate structural modulation that has been already explained by
various models. LePage et al[19] solved this structural problem,
using chemical substitutions of copper for other 3d metals. We
here present a summary of the structure of $Bi_2Sr_3Fe_2O_{9+y}$ that
explains how the modulation is related to the presence of interstitial oxygens oxygens and thereafter to a mixed valency of copper
formally above 2.

Synthesis procedures for the substitution of cooper by Co, Fe
and Mn in the $Bi_2Sr_2Ca_{n-1}Cu_nO_y$ are given elsewhere[20]. In the case
of the n=2 phase (85K superconductor) and Fe substitution, the Ca
planes must be replaced by SrO planes in the same time that divalent five-fold coordinated Cu is substituted by trivalent six-fold
coordinated Fe (Fig. 2a.).

X-ray single crystal studies[19] find that $Bi_2Sr_3Fe_2O_{9+y}$ is
isostructural to the 85K superconductor $Bi_2Sr_2Ca_1Cu_2O_8$ with an
orthorhombic structure (space group B222). However, the unit cell
parameters are a=27.245, b=5.426, c=31.696 angstroms. The a axis
is five times the basic perovskite unit. This is explained by the
fact that the modulation is, in this case, commensurate.

As already emphasized[6], the origin of the modulation is the misfit between the ionic radius of Bi that is too small to accommodate the Bi-O distance of a rocksalt-type structure and too large for a bridging conformation such as shown by Fig. 2b. A more stable configuration is then favorized by the distortion of the BiO planes thus creating zones of compression or zones of elongation (Fig. 2c) that favorize either the rocksalt or the bridging type of bonding. Fig. 3d shows projections of the BiO planes in the xy and xz planes of the cell. The displacements created by the distortion of the structure are so large that they can be easily observed together with structural modulation. The important feature is, in fact, the insertion of one extra row of oxygen every 10 Bi rows.

By applying theses results to the 85K superconducting cuprate which has a incommensurate but similar modulation, a composition $Bi_2Sr_2Ca_1Cu_2O_{8.2}$ is found (y=.2) in good agreement with the value of y=0.15 determined from the chemical analysis[4]. This should correspond to a copper formal valency of 2.2 but care must be taken because of the other possible disorders such as cationic vacancies and substitutional defects. For example, there has been reports that the Tc and amount of charge carriers are increased when the phase has been annealed under argon while the modulation remains unchanged[21]. This may be explained by the loss of the volatile Bi that could occur upon annealing.

DISCUSSION

We have provided here evidence for the role of local defects on the charge balance. This shows that both the oxygen stoichiometry and the configuration of oxygen play an important role in the electronic structure of these materials. Structural defects may affect the charge balance as well as charge transfer between planes (case of Co doping in $YBa_2Cu_3O_7$) and then provide the non-stoichiometry necessary for the mixed valency of copper (case of the Bi phases).

ACKNOWLEDGMENTS

We thank B.J. Bagley and J.H. Wernick for helpful discussions.

REFERENCES

1. R.J. Cava, B. Batlogg, C.H. Chen, E.A. Rietman, S.M. Zahurak and D. Werder, Nature 329, 423, (1987); Phys. Rev. B36 5717 (1987).
2. P.K. Gallagher, J.M. O'Bryan, S.A. Sunshine, and D.W. Murphy, Mat. Res. Bull. 22, 995 (1987).
3. R.J. Birgeneau and G. Shirane, in Physical Properties of high Temperatures Superconductors, ed. D.M. Ginsberg, World Scientific Publishing (1989).

4. J.M. Tarascon, Y. LePage, P. Barboux, B.G. Bagley, L.M. Greene,
    W.R. McKinnon, G.W. Hull, M. Giround and D.M. Hwang, Phys.
    Rev. B37, 9382 (1988).
5. J.M. Tarascon, P. Barboux, G.W. Hull, R. Ramesh, L.M. Greene,
    M. Giround, M.S. Hegde and W.R. McKinnon, Phys. Rev. B39, 4316
    (1989).
6. R.J. Cava et al. Physica C 153-155, 560 (1988).
7. D.E. Aspnes and A. Studna, Appl. Opt. 14, 220 (1975).
8. M.K. Kelly, P. Barboux, J.M. Tarascon, D.E. Aspnes, W.A.
    Bonner and P.A. Morris, Phys. Rev. B38, 870 (1988).
9. M.K. Kelly, P. Barboux, J.M. Tarascon, D.E. Aspnes, submitted.
10. M. Garriga, J. Humlicek, M. Cardona and E. Schonherr, Solid
    State Comm. 66, 1231 (1988).
11. J.M. Sanchez, F. Mejía-Lira and J.M. Morán-López, Phys. Rev.
    B, 37, 3678 (1988).
12. W.R. McKinnon, M.L. Post, L.S. Selwyn and G. Pleizier, J.M.
    Tarascon, P. Barboux, L.H. Greene, and G.W. Hull, Phys. Rev.
    B38, 6543 (1988).
13. J.M. Tarascon, P. Barboux, P.F. Miceli, L.H. Greene, G.W.
    Hull, M. Eipschutz and S.A. Sunshine, Phys. Rev. B 37, 7458
    (1988).
14. J.M. Tranquada, S.M. Heald, A.R. Moodenbaugh and Y. Xu, Phys.
    Rev. B 38, 8893 (1988).
15. J.W. Lynn, W.H. Li, H.A. Mook, B.C. Sales, Z. Fisk, Phys. Rev.
    Lett., 60, 2781 (1988).
16. P.F. Miceli, J.M. Tarascon, L.H. Greene, P. Barboux, M. Giroud,
    D.A. Neumann, J.J. Rhyne, L.F. Schneemeyer and J.V. Waszczak,
    Phys. Rev. B38, 9209 (1988).
17. P.F. Miceli, J.M. Tarascon, L.H. Greene, P. Barboux, J.D.
    Jorgensen, J.J. Rhyne and D. Neumann, Proc. MRS Symposium, San
    Diego 1989.
18. J. Clayhold, N.P. Ong, Z.Z. Wang, J.M. Tarascon and P.Barboux,
    Phys. Rev. B. 39 7324 (1989).
19. Y. LePage, W.R. McKinnon, J.M. Tarascon and P. Barboux, Phys.
    Rev.B (in press).
20. J.M. Tarascon, P.F. Miceli, P. Barboux, D.M. Hwang, G.W. Hull,
    M. Giround, Y. LePage, W.R. McKinnon, E. Tselepis, G.Pleizier,
    M. Eibschutz, D.A. Neumann and J.J. Rhyne, Phys. Rev. B in
    press.
21. D.E. Morris, C.J. Hultgren, A.M. Markelz, J.Y.T. Wei, N.G.
    Asmar and J.H. Nickel, Phys. Rev. B 39, 6612 (1989).

EXPERIMENTAL STUDY OF HIGH TEMPERATURE SUPERCONDUCTORS THROUGH SUBSTITUTION

C.L. Chien, Gang Xiao, and Marta Z. Cieplak

Department of Physics and Astronomy
The Johns Hopkins University, Baltimore, Md 21218, USA

We have studied the cuprate superconductors by the means of cation doping using different elements, inducing changes in structure, oxygen ordering, carrier concentration, magnetic properties, and indeed the superconducting properties. The response of various properties to doping allows us to elucidate the role of a particular constituent of a superconductor. We will review the results of substitution for the Cu sites in $YBa_2Cu_3O_7$. New results of cation substitution for the La and Cu sites in $La_{1.85}Sr_{0.15}CuO_4$ will be presented and discussed.

1. Introduction

Since the discoveries of $LiTi_2O_4$ and $Ba(Pb,Bi)O_3$ with $T_c \sim 13$ K in the early 1970's, oxide superconductors remained in obscurity until 1986 when superconductivity near 30 K in La-Ba-Cu-O was discovered. In a short span of just two years, one witnessed unprecedented activities in the field of high T_c superconductors. In quick succession $(La-Sr)_2CuO_4$, $YBa_2Cu_3O_7$, Bi-Sr-Ca-Cu-O, Tl-Ba-Ca-Cu-O, $(Ba_{1-x}K_x)BiO_3$, and $(Nd-Ce)_2CuO_4$ superconductors were discovered, elevating T_c to 125 K.

Despite these advances, the new physics revealed by the high T_c superconductors is not well understood. There have been a number of models proposed for the superconducting mechanism and the anomalous normal-state properties, yet none has been confirmed experimentally. Many approaches have been taken towards the understanding of this new class of materials. Probing the superconductors by both magnetic and non-magnetic impurities is one such approach. Substitution study has been proven very fruitful and effective in the studies of both conventional low-temperature superconductors and high T_c superconductors. Substitution permits the identification of active components and assemblies in the structure, the alteration of carrier concentration, the investigation of enhancement and suppression of T_c, the study of pair-breaking effects, and the correlation between superconductivity and normal-state properties.

With the exception of (Ba–K)BiO$_3$, all the other high T$_c$ superconductors are highly anisotropic, quasi two-dimensional cuprate superconductors[1]. Of those, La$_{2-x}$Sr$_x$CuO$_4$ (2-1-4 compound) and YBa$_2$Cu$_3$O$_7$ (1-2-3 compound) are distinct. These two compounds can be readily made into single phase materials without the complication of multiphase formation and the intergrowth of different layer stacking as in the Bi- and Tl-based superconductors. These are essential prerequisites for all substitution studies. The parent compounds, La$_2$CuO$_4$ and YBa$_2$Cu$_3$O$_6$, which are insulating and antiferromagnetic are already known and extensively studied[2]. The crucial Cu-O assemblies are also the simplest in these two cases, La$_2$CuO$_4$ having a single Cu–O$_2$ plane[3], and YBa$_2$Cu$_3$O$_7$ having both Cu–O$_2$ planes and Cu-O chains[4].

In this work, we will describe our results of substitution at the cation sites of the 2-1-4 and 1-2-3 superconductors. We will first briefly review the results of the substitution for the Cu sites in the 1-2-3 compound. Special attention will be paid to the substitution study of the 2-1-4 compound, where the role of the Cu-O$_2$ can be assessed without the complication of the Cu-O chains or other more complex structures. Effects of substitution on superconducting and normal-state properties will be presented. We will investigate to what extent impurity-induced disorder and local moments affect high T$_c$ superconductivity. Finally, we discuss the results of (La–Gd)$_{1.85}$Sr$_{0.15}$CuO$_4$ where the La-site has been substituted by Gd. All three phases (T, T′, and T*) of the 2-1-4 compound have been realized in this system. Their magnetic, conducting and superconducting properties will be discussed.

2.   Results of Cu Substitution in 1-2-3 Compound

Before discussing the Cu substitution results of the 2-1-4 compound, it is instructive to review the substitution results of the 1-2-3 compound. Shortly after the discovery of high T$_c$ YBa$_2$Cu$_3$O$_7$, it was found that the Y-site is electronically inactive and can be completely replaced by essentially all 3+ rare earth (RE) elements[5]. T$_c$ remains close to 90 K while the RE ions retain their magnetic moments and order antiferromagnetically at low temperatures. The Cu-sites however are crucial as first demonstrated by the substitution work of Xiao et al.[6] and Maeno et al.[7] However, since there are two Cu sites, one must determine the resulting crystal structure due to doping, the site location of the substituting atoms, and the occupations of all the oxygen sites through x-ray and neutron diffractions.

The variation of T$_c$ as a result of Cu substitution are shown in Fig. 1. Neutron diffraction data show that Zn occupies only the Cu(2)-site (plane-site), whereas Al, Co, Ga occupy only the Cu(1)-site (chain-site)[8,9]. When the Cu(2) site is substituted, as in the case of Zn, the orthorhombic structure, the linear chain structure and the accompanying oxygen vacancy order remain unchanged. When the Cu(1) site is substituted by Al, Co and Ga, the chain structure is quickly disrupted and the structure becomes tetragonal. When the Cu(1) site is substituted, T$_c$ degrades slightly and smoothly across the orthorhombic-tetragonal transition, whereas the occupancy of the Cu(2) site by Zn rapidly suppresses T$_c$[10]. Thus in the 1-2-3 compound the integrity of the Cu-O$_2$ planes is essential for high T$_c$ superconductivity. It is interesting to note that the preferential substitution of the two Cu sites is apparently related to the charge state of the dopant. All the Cu(1) dopants (Co, Al, Ga) are in 3+ charge state, drawing a small amount of additional oxygens into the structure, resulting a total oxygen content higher than 7. The Cu(2) dopant (Zn) has a 2+ state same as that of Cu^{2+} without changing the total oxygen content. Co^{3+}, differing from Al^{3+} and Ga^{3+}, carries a localized moment thus suppresses T$_c$ more strongly.

$Cu^{2+}(3d^9)$ is a Jahn-Teller ion with a magnetic moment associated with its spin $S = \frac{1}{2}$. In the insulating parent compound the Cu moments order antiferromagnetically. Upon doping with holes, the three-dimensional antiferromagnetic ordering is rapidly degraded, but the 2D antiferromagnetic ordering persists. The half-filled $3d_{x^2-y^2}$ states of the $Cu^{2+}(3d^9)$ and the oxygen p orbitals are believed to be crucial for high T_c superconductivity. Zn^{2+} and Ga^{3+} are two unique dopants because both have a $3d^{10}$ configuration. Upon doping of Zn, the 3d hole at the substituted Cu site is eliminated. Thus hopping of the mobile holes onto the substituted site with a full 3d configuration become energetically unfavorable. This accounts for the strong suppression of T_c.

3. Results of Cu Substitution in 2-1-4 Compound

After establishing the $Cu-O_2$ plane being the important assembly of the 1-2-3 compound, it is most fruitful to study the doping effect of the 2-1-4 compound where there is only one $Cu-O_2$ plane. Here we shall discuss the results of Zn-doped and Ga-doped samples. The crystal

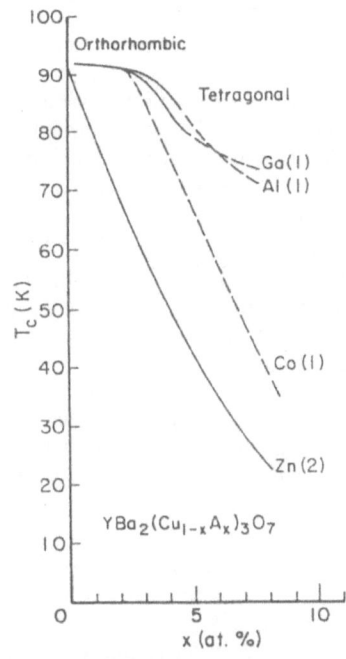

Fig.1 Variation of T_c with the dopant content for 1-2-3 compound doped with Zn, Co, Al and Ga.

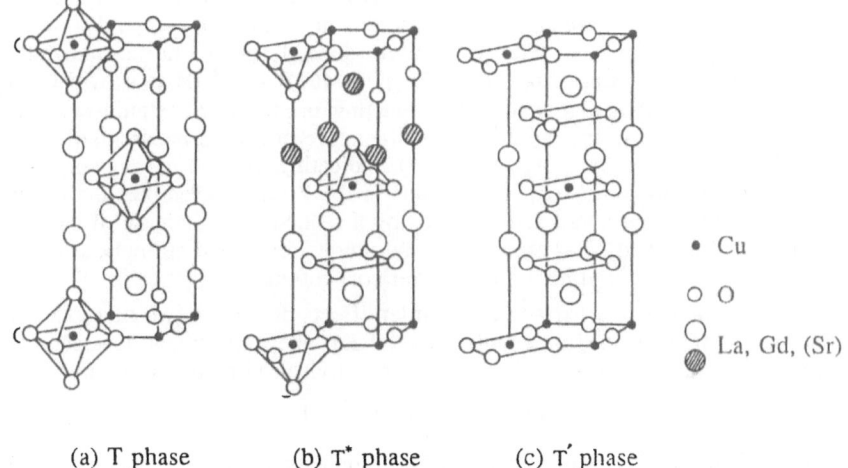

(a) T phase   (b) T^* phase   (c) T' phase

Fig.2 Structures of (a) the T phase: $La_{1.85}Sr_{0.15}CuO_4$, (b) the T^* phase: $(La_{0.55}Gd_{0.45})_{1.85}Sr_{0.15}CuO_4$, and (c) the T' phase: $Gd_{1.85}Sr_{0.15}CuO_4$.

structure of La_2CuO_4, shown in Fig. 2(a), also known as the T-phase, contains the 2D planes of Cu-O octahedrons. The substitution of Sr^{2+} for the La^{3+} site in $La_{2-x}Sr_xCuO_4$ creates free holes with concentration (p) increasing with x (Ref.11). Upon Sr doping the Néel temperature (T_N) of La_2CuO_4 decreases very rapidly. An orthorhombic to tetragonal structural transition is induced as well. When sufficient amounts of mobile holes are present in the $Cu-O_2$ plane, superconductivity appears with T_c increasing with p, reaching a maximum of 38K at x=0.15. Further increase in p lowers T_c. Beyond x=0.32, the material becomes a better metal but ceases to be superconducting. This is one of the puzzling features of the 2-1-4 compound.

The unique features ($3d^{10}$ configuration) of Zn^{2+} and Ga^{3+} have already been emphasized. Their sharply different effects in the 1-2-3 compound are due to the different substituting sites. In the 2-1-4 compound, both of them can only occupy the plane sites. However, a number of similarities and differences can be anticipated in the Zn-doped and Ga-doped samples. Because of the $3d^{10}$ configuration, both Zn^{2+} and Ga^{3+} are not Jahn-Teller ions, and carry no magnetic moment. The carrier concentration of the Zn-doped samples is likely to remain unchanged because Zn^{2+} and Cu^{2+} have the same charge state, whereas the Ga-doped samples where Ga is in the 3+ charge state are expected to lead to reduced hole concentrations.

The most dramatic effect due to both Zn and Ga is the strong suppression of T_c[12,13]. A mere 2.5% of either Zn or Ga reduces T_c to zero as shown in Fig. 3. The striking similarity of the Zn-doped and Ga-doped samples is evident. Zn^{2+} and Ga^{3+} are both effective in suppressing T_c when they occupy the plane sites. This is distinctively different from that of the 1-2-3 compound. We believe that the substitution study of Zn and Ga is likely to elucidate the pairing mechanism of high T_c superconductivity, which so far remains elusive. In conventional s-wave superconductors, low doping level of non-magnetic impurities have no effect on T_c. On the other hand, magnetic impurities have a large effect, because the spin-flip scattering caused by the magnetic impurities is strongly pair-breaking. In the high T_c 2-1-4 compound, both nominally non-magnetic Zn and Ga impurities have strong suppressing effect on T_c.

To provide further insight into the 2-1-4 compound doped with Zn and Ga, we resort to resistivity and susceptibility measurements which respectively probe the disorder and magnetic effects. The temperature dependence of the normalized resistivity $[\rho(T)/\rho(297\ K)]$ of $La_{1.85}Sr_{0.15}(Cu_{1-x}Zn_x)O_4$ with $0 \le x \le 0.04$ is shown in Fig. 4. All the samples, including the non-superconducting ones, show metallic behavior. It should be noted that the resistivity at room temperature $\rho(297\ K)$ is essentially constant and insensitive to Zn content (inset of Fig. 4), indicating that the carrier concentration is unaffected by Zn doping as expected. Thus in the Zn-doped samples the disappearance of superconductivity is not accompanied by a reduction or increase of carrier concentration. The Zn-doped 2-1-4 system has the unique feature of strongly affecting the pairing mechanism without affecting the carrier concentration.

To more quantitatively extract the disorder effects due to Zn, we first examine the resistivity in the high temperature range (T>200 K) where the linear T dependence is evident (Fig. 4). In this temperature range the resistivity can be described by

$$\rho(T) = \rho_0 + \rho(x) + \alpha(x)T , \qquad (1)$$

where ρ_0 is the residual resistivity due to non-Zn defects, $\rho(x)$ is the residual resistivity due to impurity scattering induced by Zn, $\alpha(x)T$ is the linear T term with a slope of $\alpha(x)$. The residual resistivity at T = 0 K,

$$\rho(0\,K) = \rho_0 + \rho(x) , \qquad (2)$$

can be obtained by extrapolating the high temperature data to $T = 0$ K. The value of $\rho(0\,K)$ so obtained varies linearly with the Zn content as shown in Fig. 5 indicating that the Zn impurities are uniformly distributed in the samples.

The second term in Eq. (1), due to impurity scattering induced by Zn, can be written as[14]

$$\rho(x) = \frac{4\pi v_F}{\omega_p^2 l} , \qquad (3)$$

where v_F is the Fermi velocity, $\omega_p = (\frac{4\pi n e^2}{m^*})^{1/2}$ is the plasma frequency, and l is the mean free path associated with impurity scattering. Using $v_F \sim 0.95 \times 10^7$ cm/sec and $\hbar\omega_p \sim 0.6$ eV, determined for the 2-1-4 compound[15], we have calculated l according to Eq. (3). Fig. 6 shows T_c as a function of $\rho(x)$ and l. T_c is reduced to zero when l is about 55 A which is still considerably larger than the coherence length (~ 16 A) This analysis suggests that effects of disorder alone can not satisfactorily account for the T_c suppression in Zn-doped 2-1-4 compound.

We next turn to the results of Ga-doped 2-1-4 compound in the form of $La_{1.85}Sr_{0.15}(Cu_{1-x}Ga_x)O_4$. The temperature dependence of the resistivity is shown in Fig. 7. First of all, one notes that the room temperature resistivity $\rho(297\,K)$ increases with Ga content. This is because when Cu^{2+} is substituted by Ga^{3+}, each Ga^{3+} would ideally remove one free hole. The reduced hole concentration causes a higher resis-

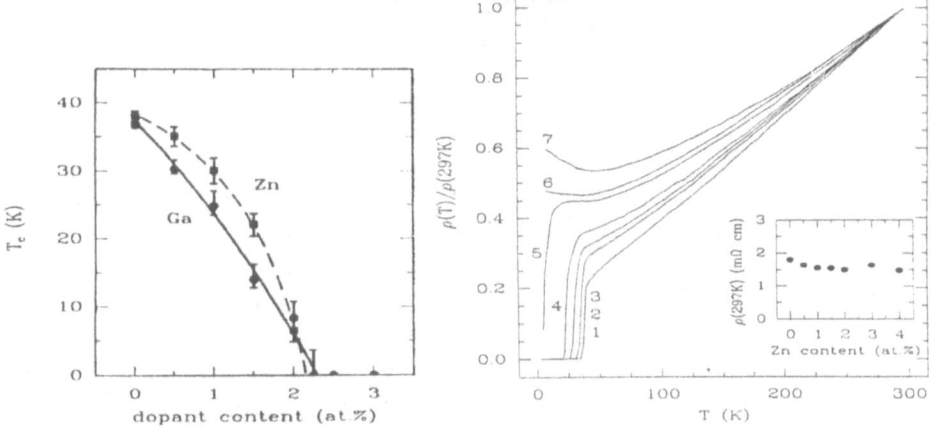

Fig.3 Variation of T_c with the dopant content in Zn-doped (squares) and Ga-doped (circles) 2-1-4 compound. The bars represent 90-10% resistive superconducting transition.

Fig.4 The temperature dependence of normalized resistivity for samples with (1) x=0, (2) 0.005, (3) 0.01, (4) 0.015, (5) 0.02, (6) 0.03, and (7) 0.04. Inset: $\rho(297\,K)$ vs Zn content.

tivity. Quantitatively, the increase in $\rho(297\ K)$ is somewhat larger than that expected for one hole reduction for each Ga^{3+} substitution. The residual resistivity and mean free path due to Ga impurity scattering have been calculated from the data, and are shown in Fig. 6. They are very similar to those of the Zn-doped samples. These residual resistivity analyses show that in both the Zn-doped and Ga-doped samples, the strong T_c suppression effect can not be accounted by disorder alone.

Next we discuss the magnetic susceptibility data presented in Fig. 8. The magnetic susceptibility (χ) of the superconducting $La_{1.85}Sr_{0.15}CuO_4$ is slightly temperature-

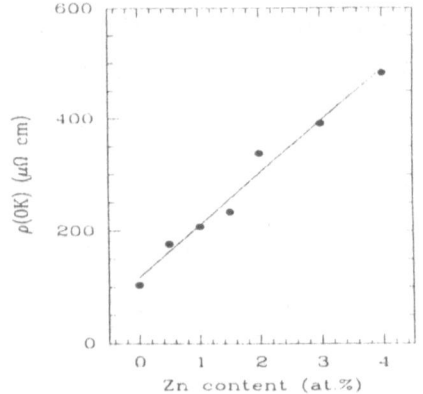

Fig.5 Residual resistivity as a function of Zn content.

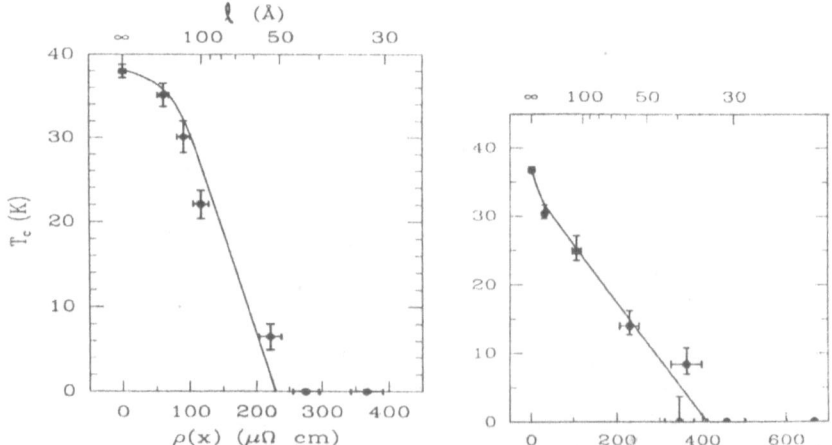

Fig.6 T_c versus $\rho(x)$, the residual resistivity due to Zn-induced disorder (top) and Ga-induced induced disorder (bottom). The upper scale is the mean-free path due to disorder, l.

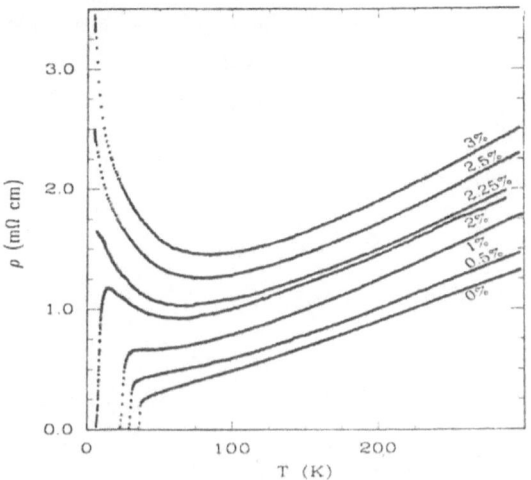

Fig.7 The temperature dependence of resistivity for samples with various Ga content (in at.%).

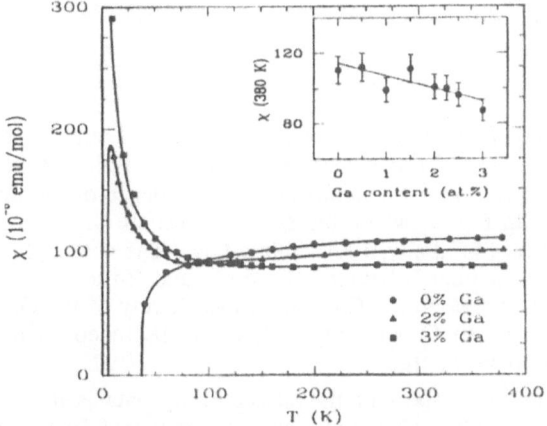

Fig.8 The temperature dependence of the susceptibility for samples with Ga content of 0, 2, and 3 at.%. The inset shows χ(380 K) as a function of the doping level.

dependent due to antiferromagnetic fluctuation in the Cu-O$_2$ planes. With Ga doping, the susceptibility at high temperature i.e. χ(380 K) gradually decreases as shown in the inset of Fig. 8, due to a slight lowering of hole concentration. Most interestingly, at low temperature, the Ga-doped samples exhibit a Curie-Weiss behavior,

$$\chi = \chi_0 + \frac{N p_{eff}^2 \mu_B^2}{3 k_B (T - \theta)} , \tag{5}$$

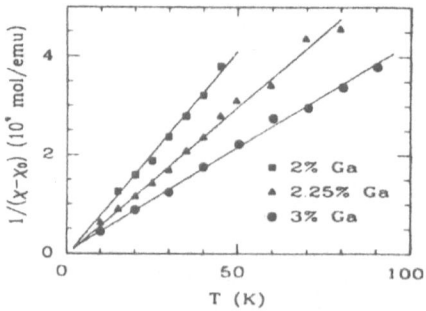

Fig.9 The inverse of the Curie-Weiss susceptibility, $\chi - \chi_0$, versus temperature for 3 samples: 2, 2.25 and 3 at.% of Ga content.

where χ_0 is the temperature independent part of the susceptibility, N is the number of magnetic ions, p_{eff} is the effective moment in units of Bohr magneton (μ_B), and θ is the Curie-Weiss temperature. The temperature dependence of the inverse susceptibility $1/(\chi - \chi_0)$ is shown in Fig. 9, where the straight lines are the best-fit results using Eq. (5). In all cases θ is essentially 0 K within ±1 K. The susceptibility data of the Zn-doped samples show similar behavior. Since Ga and Zn are non-magnetic, the magnetic moment has to reside on the Cu-sites in the vicinity of the Ga and Zn impurities. The magnetic moment per Ga impurity is found to be independent of the Ga content and has a value of about 0.8 μ_B.

The most important aspect of the susceptibility data is that Ga and Zn dopants induce localized moments. This may seem surprising at first since Ga^{3+}(3d^{10}) and Zn^{2+}(3d^{10}) are both non-magnetic ions. However, one recalls that in the superconducting 2-1-4 compound, the spin S = ½ of the Cu^{2+} ions do not reveal themselves because of antiferromagnetic fluctuation. When a nominally non-magnetic Ga^{3+} and Zn^{2+} substitutes Cu^{2+}(3d^9), the spin S = ½ associated with that Cu site will be removed. There is, then, a net spin ½ now localized in the vicinity of the Ga or Zn site. It is with this scheme that the substitution of Ga or Zn generates localized moments. This is a peculiar consequence of the antiferromagnetic background of the Cu-O$_2$ planes. The appearance of localized moments on the crucial Cu-O$_2$ planes, as induced by Ga^{3+} and

Zn^{2+}, raises the distinct possibility that the strong T_c suppression may be of magnetic origin, caused by a magnetic pair-breaking mechanism.

4. Results of La Substitution of 2-1-4 Compound

La_2CuO_4 has the well known structure (T-phase), with a plane of Cu-O octahedrons as shown in Fig. 2(a). In addition, there are two other closely related structures containing the $Cu-O_2$ plane. For a series of compounds Ln_2CuO_4, where Ln is a light rare earth (Pr to Gd), the Cu atom is square-planer coordinated (T'-phase) as shown in Fig. 2(c). The T'-phase has attracted a great deal of attention because of the recent discovery of electron superconductor in $(Nd-Ce)_2CuO_4$ (Ref.16). Finally there is the T*-phase (Fig. 2(b)) where the Cu atom is pyramidally coordinated. The positions of the copper and the rare earth atoms remain identical in the T, T' and T* phases, only the positions of the oxygen atoms vary. Because of the close similarity among the T, T' and T* phases, and the occurrence of both hole and electron superconductivity, it would be very desirable if a system can be realized, in which all three phases exist. This has been accomplished in $(La_{1-x}Gd_x)_{1.85}Sr_{0.15}CuO_4$ ($0 \leq x \leq 1$) (Ref.17).

We have fabricated a large number of samples of $(La_{1-x}Gd_x)_{1.85}Sr_{0.15}CuO_4$. A rather complete phase diagram has been determined as shown in Fig.10. In the La-rich end, pure T-phase has been found with $0 \leq x \leq 0.1$. In the Gd-rich end, pure T'-phase is realized with $0.95 \leq x \leq 1$. The T*-phase exists in the composition range of $0.42 \leq x \leq 0.49$. The x-ray diffraction pattern of the three phases are shown in Fig. 11. The samples are of very high quality that there are no detectable impurity phases and every peak can be indexed. Between the phase boundaries of the three phases, mixed phases co-exist. In $0.1 \leq x \leq 0.42$, both T and T* phases are found. The amount of T*-phase increases steadily at the expense of T-phase as shown in Fig.10. Similarly, in $0.55 \leq x \leq 0.95$, both T* and T'-phases are found.

Magnetic measurements show that the T-phase and the T*-phase samples are paramagnetic. The magnetic susceptibility can be well described by the Curie-Weiss law of Eq. (5). Fig.12 shows the inverse of susceptibility as a function of temperature for various Gd-doped T and T* phase samples. The solid lines are fits to the data using relation (5). The fitted parameter θ is about 0 ± 1 K for $(La_{1-x}Gd_x)_{1.85}Sr_{0.15}CuO_4$ with $0 \leq x \leq 0.1$, indicating that the magnetic interaction between the doped Gd ions is negligibly small. However, in $(La_{0.5}Gd_{0.5})_{1.85}Sr_{0.15}CuO_4$, the obtained θ is -9.1 K, a signature of the existence of anti-ferromagnetic interaction among the Gd ions. The obtained magnetic moment p_{eff} of the Gd^{3+} ions is shown in the inset of Fig.12 for various samples. The value p_{eff} remains constant for both T and T* phases with a value of 7.95, which is identical to $g\sqrt{J(J+1)} = 7.94$, the Gd^{3+} ground state moment according to Hund's rules.

On the other hand, T'-phase samples of both Gd_2CuO_4 and $Gd_{1.85}Sr_{0.15}CuO_4$ exhibit very interesting magnetic orderings. The temperature dependence of suscepti-bilities of the two samples is shown in Fig.13. As a comparison, the data of $(La_{0.5}Gd_{0.5})_{1.85}Sr_{0.15}CuO_4$ (T*-phase) is also included. There exist two magnetic transi-tions at approximately $T_N=285$ K and 20 K in the T' phase. The transition at $T_N=285$ K is undoubtedly associated with the antiferromagnetic ordering of the Cu-O sublattice, similar to the one seen in the La_2CuO_4 with a Néel temperature $T_N = 260$ K. The nature of the low temperature magnetic transition, however, is less clear, prob-ably involves some sort of ordering of the Gd moments.

In La_2CuO_4 (T-phase), the Néel state of the Cu sublattice is extremely susceptible to Sr doping onto the La sites. T_N is suppressed to zero with less than 1 at.% of Sr. At sufficiently high carrier concentration the samples become superconducting. For the Gd_2CuO_4 (T'-phase) the situation is completely different. The value of $T_N = 285$ K is insensitive to Sr doping at all as shown in Fig. 13. This indicates that Sr doping does not introduce extra mobile holes in the $Cu-O_2$ plane. Instead, these holes are trapped in the vicinity of the local Sr sites. However, the T'-phase can be doped with electrons by substituting Ln^{3+} with Ce^{4+} to become the electron superconductor. An examination of the structures shown in Fig. 2 suggests that the apical oxygen atoms play a key role. Doping of holes or electrons on the Cu-O planes depends strongly on the local geometry of the Cu-O polyhedrons.

We have also studied the superconductivity in the T-phase of $(La_{1-x}Gd_x)_{1.85}Sr_{0.15}CuO_4$ ($x \leq 0.1$). The results are shown in Fig. 14. When La is sub-stituted by Gd, T_c decreases linearly from $T_c = 38$ K. The T_c suppression effect by Gd is in sharp contrast to the ineffectiveness of rare earths on T_c in the 1-2-3 system. At $x = 0.1$ there is a clear break of the slope because at $x > 0.1$, the T* phase appears.

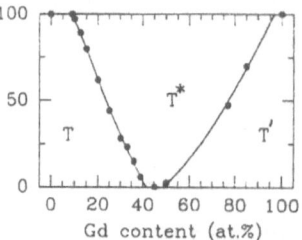

Fig.10 Structural phase diagram of $(La_{1-x}Gd_x)_{1.85}Sr_{0.15}CuO_4$.

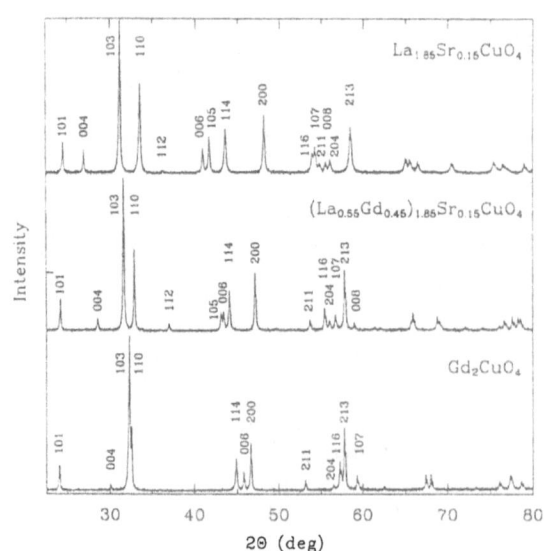

Fig.11 θ–2θ x-ray powder diffraction patterns of the T phase $La_{1.85}Sr_{0.15}CuO_4$, the T* phase $(La_{0.55}Gd_{0.45})_{1.85}Sr_{0.15}CuO_4$, and the T' phase Gd_2CuO_4.

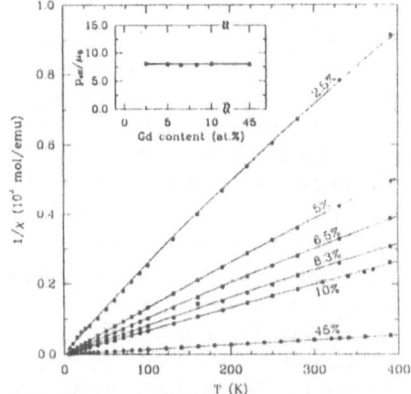

Fig.12 Temperature dependence of the inverse magnetic susceptibility for samples with the T and T* phases. The solid lines are the fits to the data using the Curie-Weiss law. The inset shows the effective moment obtained from the fits versus Gd content.

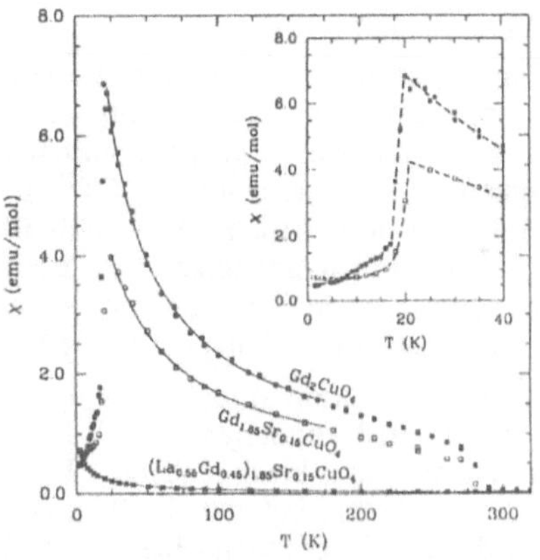

Fig.13 Temperature dependence of magnetic susceptibility for T' phase samples Gd_2CuO_4 and $Gd_{1.85}Sr_{0.15}CuO_4$. The solid lines are the fits to the data using Curie-Weiss law. Data for $(La_{0.55}Gd_{0.45})_{1.85}Sr_{0.15}CuO_4$ are included for comparison. The inset shows susceptibility at low temperatures.

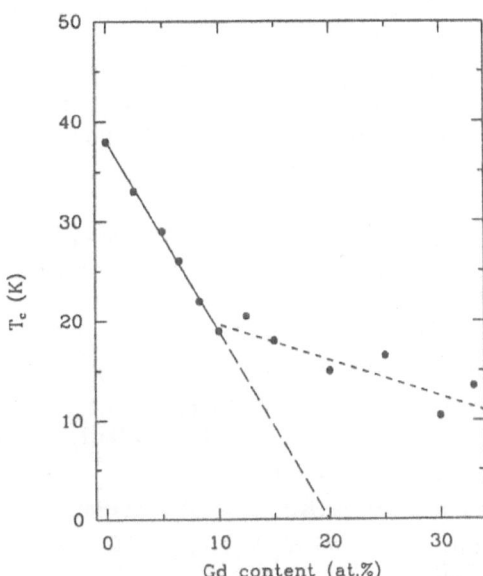

Fig.14 Superconducting transition temperature T_c versus Gd content. The critical concentration obtained from extrapolation is $x_c = 0.2$.

5. Conclusions

Substitution studies of the 2-1-4 compound reveal a wealth of information and new properties. The Cu-O_2 plane, common to all high T_c cuprate superconductors, plays a pivotal role. When nominally non-magnetic Zn^{2+} and Ga^{3+}, having $3d^{10}$ configuration, substitute for the Cu^{2+} sites, T_c is sharply reduced to zero at a doping concentration of less than 2.5%. This illustrates the crucial role of 3d holes at the Cu-site for high T_c superconductivity. Because of the unusual circumstance of the Cu-O_2 plane, non-magnetic Ga^{3+} and Zn^{2+} induce localized moments near the dopant sites. It raises the possibility of magnetic pair breaking effect for the strong suppression of T_c.

The substitution study of La by Gd in the 2-1-4 compound is equally profitable. All three related structures (T, T' and T*) differing only in the local copper-oxygen environments have been realized. It has been found that the T-phase can be doped with mobile holes, but not the T'-phase. Unlike the 1-2-3 compound, the rare earth site in the 2-1-4 compound has a significant influence on the superconducting properties.

References

(1)   M. Oda, *et al.*, Phys. Rev. B **38**, 252 (1988), and *references therein*.

(2)   D. Vaknin *et al.*, Phys. Rev. Lett. **58**, 2802 (1987); G. Shirane *et al.*, *ibid.* **59**, 1613 (1987); Y. Endoh *et al.*, Phys. Rev. B **37**, 7443 (1988); J. M. Tranquada *et al.*, Phys. Rev. Lett. **60**, 156 (1988).

(3)   J. M. Longo and P. M. Raccah, J. Solid State Chem. **52**, 254 (1984); K. K. Singh *et al.*, *ibid.* **52**, 254 (1984).

(4)   J. D. Jorgensen *et al.*, Phys. Rev. B**36**, 3608 (1987);

(5)   D. W. Murphy *et al.*, Phys. Rev. Lett. **58**, 1888 (1987); P. H. Hor *et al.*, *ibid.* **58**, 1891 (1987); Gang Xiao *et al.*, Solid St. Commun. **63**, 817 (1987).

(6)   Gang Xiao, *et al*, Phys. Rev. B **35**, 8782 (1987).

(7)   Y. Maeno, *et al*, Nature **328**, 512 (1987).

(8)   Gang Xiao, *et al*, Nature (London) **332**, 238 (1988); C.L. Chien, *et al*, in *Superconductivity and Its Applications*, edited by H.S. Kwok and D.T. Shaw (Elsevier, New York, 1988), p.110.

(9)   H. Maeda, *et al*, Physica C **157**, 483 (1989).

(10)   Gang Xiao, *et al*, Phys. Rev. Lett. **60**, 1446 (1988).

(11)   N. P. Ong *et al*.,Phys. Rev. B **35**, 8807 (1987); M. W. Shafer *et al.*, *ibid.* **36**, 4047 (1987); J. B. Torrance *et al*.,Phys. Rev. Lett.**61**, 1127 (1988).

(12)   Gang Xiao, *et al*, Phys. Rev. B **39**, 315 (1989).

(13)   Marta Z. Cieplak, *et al*, Phys. Rev. B **39**, 4222 (1989).

(14)   J. M. Ziman, *Principles of the Theory of Solids* (Cambridge University, Cambridge, 1979), p.220.

(15)   M. Gurvitch and A. Fiory, Phys. Rev. Lett. **59**, 1337 (1987).

(16)   Y. Tokura, H. Takagi, and S. Uchida, Nature **337**, 55 (1989).

(17)   Gang Xiao, Marta Z. Cieplak and C. L. Chien (to be published).

THEORETICAL STUDY OF OXYGEN DISORDER EFFECTS IN

$YBa_2(Cu_{1-x}M_x)_3O_{7+y}$ (M = Al, Co, Fe, Ga)

G. Baumgärtel and K.H. Bennemann

Institute of Theoretical Physics
Freie Universität Berlin
Arnimallee 14, D-1000 Berlin 33, W.-Germany

We study structural changes in $YBa_2Cu_3O_{7-\delta}$ resulting from alloying with Al, Co, Fe and Ga. For the various interatomic interactions we use a bond model. Then we find at T = 0 that the tetragonal phase seemingly observed by x-ray diffraction consists of differently oriented orthorhombic domains. The observed uptake of excess oxygen as a consequence of doping with Al, Co and Fe is explained by domain formation. In addition, we propose a model to treat the system at T > 0.

INTRODUCTION

The high T_c superconductor $YBa_2Cu_3O_{7-\delta}$ with a critical temperature around 90K has an orthorhombically distorted lattice that is induced by ordering of oxygen ions and vacancies in the Cu(1) planes, where Cu-O chains are formed[1]. In order to understand the relation between T_c and the oxygen content and also the ordering of oxygen atoms in the Cu(1) planes several groups studied the dependence of T_c on y in $YBa_2Cu_3O_{7-y}$.[2] Of similar interest is the dependence of T_c on alloying, e.g. substitution of Cu by metal impurities. In particular, interesting effects have been observed when Cu is replaced by Al, Co, Fe or Ga. What are the various effects caused by these impurities? Is there a redistribution of oxygen caused by the impurities? These questions are studied in the following.

X-ray diffraction experiments[3] yielded that the orthorhombic distortion decreases continuously in $YBa_2(Cu_{1-x}M_x)_3O_{7+y}$ as a function of x (M=Al,Co,Fe), and already for x = 0.03 the lattice parameters a and b become equal. At this impurity content, T_c is still between 60K and 90K. Thus, these experiments seemed to show that even in the tetragonal phase $YBa_2(Cu_{1-x}M_x)_3O_{7+y}$ is a high T_c superconductor. However, investigation of the seemingly tetragonal phase with a = b by electron microscopy[4] indicates that it consists of differently oriented orthorhombic domains so that short Cu-O chains are present in this phase.

Since Al, Co, Fe and Ga impurities occupy mainly Cu(1) sites, we shall study in the following the redistribution of oxygen atoms in the Cu(1) planes as a consequence of allying using a bond model for the interatomic interactions between the oxygen and metal atoms. Our results may provide additional information about the relationship between high T_c superdonductivity in $YBa_2(Cu_{1-x}M_x)_3O_{7+y}$ and the existence of Cu-O chains in these compounds.

RESULTS AT T=0

At T=0 the driving force behind the oxygen redistribution in the Cu(1) plane of $YBa_2(Cu_{1-x}M_x)_3O_{7+y}$ is the fact that the impurities prefer a higher oxygen coordination than copper[5]. We estimate the oxygen coordination of the impurities from the strengths of the various cation-oxygen bonds using the bond-valence method[6] (which states that the sum of the bond strengths at each atom is equal to the atomic valence) and an empirical bond strength (s) - bond length (R) formula of the form[6] $s=\exp\{-f(R)\}$. We find that Co^{3+}, Fe^{3+} and Ga^{3+} impurities have a fivefold oxygen coordination, whereas Al^{3+}, Co^{4+} and Fe^{4+} ions should have six nearest oxygen neighbours. Note, the copper atoms in the Cu-O chains have only a fourfold coordination. Thus, if we substitute Al, Co, Fe or Ga for Cu, these impurities tend to attract additional oxygen atoms in order to increase their oxygen coordination in the Cu(1) plane.

In order to increase the oxygen coordination of an impurity either copper looses a nearest neighbour oxygen or extra oxygen atoms have to be put into the Cu(1) plane. Note, the uptake of excess oxygen is limited by the condition that a copper atom on a Cu(1) site should not have more than four nearest oxygen neighbours, i.e. it must not have more than two nearest oxygen neighbours with the Cu(1) plane.

Apparently, the presence of impurities on Cu(1) sites causes a redistribution of oxygen atoms in the neighbourhood of the impurities. This local oxygen redistribution around the impurities may induce a global oxygen rearragement in the whole Cu(1) plane. We introduce the attractive bond energies ε_{CuO} and ε_{MO} between nearest neighbours and the repulsive nearest-neighbour and next-nearest-neighbour interaction energies ε_{OO} and ε_{OO}', see Fig. 1. Then, the total energy contribution

$$E(p) = \sum_{i=1}^{2N} t_i (2(1-p)\varepsilon_{CuO} + 2p\varepsilon_{MO}) + \sum_{\langle ij \rangle}^{nn} t_i t_j \varepsilon_{OO} + \sum_{\langle ij \rangle}^{nnn} t_i t_j \varepsilon_{OO}' \quad (1)$$

determines the energy of various oxygen configurations in the Cu(1) plane as a function of the impurity concentration p with respect to the Cu(1) sites. Here $\sum_{i=1}^{2N}$ is a sum over all oxygen sites in the Cu(1) plane, $\sum_{\langle ij \rangle}^{nn}$ runs over all nearest-neighbour oxygen sites and $\sum_{\langle ij \rangle}^{nnn}$ over all next-nearest neighbour sites. t_i is equal to 1, if the oxygen site i is occupied, and $t_i = 0$ otherwise. For simplicity, we neglect the interactions of the atoms in the Cu(1) plane with the atoms above and below. If necessary, these interactions could be included.

We consider substitutional impurities with k=6, i.e. Al^{3+}, Co^{4+} and Fe^{4+}, and compare the energies of the following phases. In phase A, we assume that no oxygen redistribution occurs if we substitute some impurities for copper. In phase B, we assume that in the surrounding of an impurity two copper atoms have lost one nearest-neighbour oxygen which is transferred into the neighbourhood of the impurity to increase its oxygen coordination to six, but apart from this the chains remain intact. In phase C, we also assume that the impurities are sixfold coordinated, but allow random occupation of the O(1) and O(5) sites by oxygen atoms which are not nearest-neighbours of an impurity. In phase D, we assume that differently oriented orthorhombic domains consisting of Cu-O chains along the a- or b-axis are formed. Each domain has two impurities on opposite domain corners so that the impurities have four nearest oxygen neighbours in the Cu(1) plane, see Fig.1.

In general one obtains for impurities with a sixfold coordination and a bond energy with oxygen, ε_{MO}, which makes the increase in the oxygen coordination of the impurity energetically favourable (Al^{3+}, Co^{4+} and Fe^{4+}), that phase B is lower in energy than phase A for all impurity concentrations p. Furthermore, neglecting the uptake of excess oxygen due to substitutional impurities we obtain from Eq. 1 that the difference in

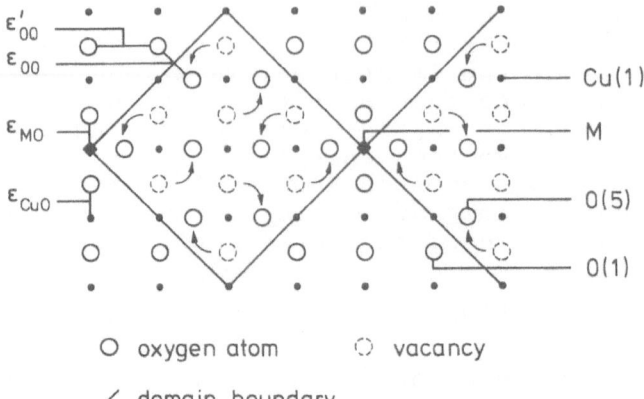

○ oxygen atom   ◌ vacancy

╱ domain boundary

Fig. 1.   Illustration fo the formation of orthorhombic
           domains induced by an impurity with a sixfold
           coordination.

energy between the phases B, C and D depends only on ε_{00}
and ε_{00}'. In particular we find[8] that the domain phase D
is lower in energy than the phase B, if

$$\left(\frac{1}{\sqrt{p}} - 2\right)\frac{\varepsilon_{00} - \varepsilon_{00}'}{\varepsilon_{00}} < 1 \tag{2}$$

It can be seen from Eq. (2) that the impurity concentration p_c, where the transition into the domain phase
occurs, depends only on the parameter $q \equiv (\varepsilon_{00} - \varepsilon_{00}')/\varepsilon_{00}$,
which is expected not to change much for different impu-
rities. In Fig. 2 results are given for the dependence
of p_c on q. Note, we estimated that allowing excess
oxygen will change p_c only slightly (less than 10%). In
addition, for impurity concentrations $p \leq 0.25$ our
calculations yield that the tetragonal phase is always
higher in energy than the phases B and D, if q>0. For
larger p the phases B,C and D become equivalent, since
at such high impurity concentrations all oxygen atoms
are fixed in the neighbourhood of an impurity. Summarizing, our results show that no tetragonal phase occurs
due to doping. Instead, local oxygen redistribution in
the neighbourhood of the impurities (phase B) is energetically favourable for $p < p_c$, whereas small orthorhombic domains are formed for $p > p_c$.

     Electron microscopy studies[4] reveal that the phase
with macroscopically equal lattice parameters a and b
(as observed by x-ray diffraction) results from an ave-

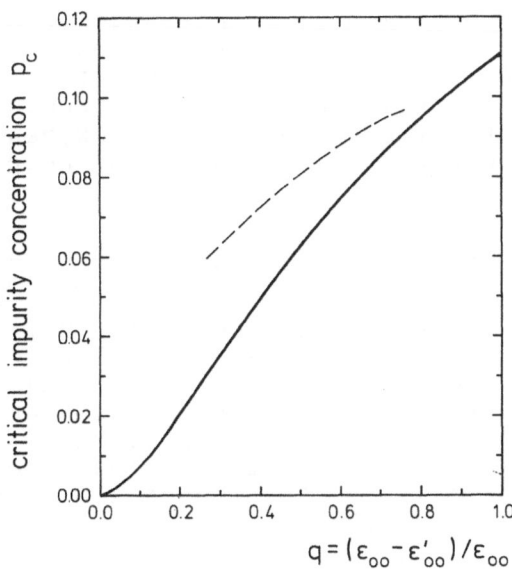

Fig. 2. Critical concentration of Impurities (p_c),
where the transition from the phase B to the
domain phase D occurs as a function of
$q \equiv (\varepsilon_{00} - \varepsilon'_{00})/\varepsilon_{00}$. Here, B refers to the
phase, where oxygen is rearranged only locally
around the impurities. Note, for impurities
with a fivefold oxygen coordination the domains
can be too large at p_c to give a = b as a
result of the averaging procedure in x-ray
diffraction. Thus, the phase with a = b is
observed experimentally at larger p as it is
qualitatively indicated by the broken line.

raging procedure over small orthorhombically distorted
domains. In x-ray experiments for $YBa_2(Cu_{1-x}M_x)_3O_{7+y}$
(M = Al, Co, Fe) the phase with a = b occurs at an impurity content x between 2.5% and 3%. This corresponds to
an impurity concentration p in the Cu(1) plane between
6% and 9%, since at least 75% of the impurities[3] seem to
occupy Cu(1) sites. Our calculated values for p_c agree
with these experimental results if $1/2 \leq q \leq 3/4$, which
is in agreement with the relation between nearest-and
next-nearest-neighbour oxygen-oxygen repulsion[9] reported
by other authors. Furthermore, the formation of orthorhombic domains gives rise to oxygen configurations in
the Cu(1) plane that cause the observed uptake of one
additional oxygen atom per impurity[8].

In an improved model, the energy $E(p)$ should be determined from an electronic theory. This may be done by using the Hubbard Hamiltonian H and a tight-binding type theory. Determining as usual the electronic Green's functions G_{ii} for the p electrons of oxygen and the d electrons of copper at site i with the help of the recursion method, then

$$E(p) = \sum_i \int_{-\infty}^{\varepsilon F} (\varepsilon - \varepsilon_i) N_i(\varepsilon) d\varepsilon \ldots - E_R - E_M + \ldots \quad (3)$$

where Σ_i runs over all occupied Cu(1), O(1) and O(5) sites in the Cu(1) plane. The local electronic density of states is given by $N_i(\varepsilon) = -(1/\pi) \text{Im} G_{ii}(\varepsilon)$. E_R and E_M refer to the Born-Mayer core-core repulsion and the Madelung Energie of the Cu(1) plane. Eq. (3) is the basic equation for an electronic calculation of the ground state energies of the phases A,B,C and D. Calculations are in progress.

FINITE TEMPERATURE PROPERTIES

It is of interest to note that the oxygen order-disorder transitions in the Cu(1) plane studied previously by various groups[9,10] are of course affected by impurities. For example, due to oxygen disorder effects in the neighbourhood of the impurities the transition from the orthorhombic to the tetragonal phase that is driven by entropy may occur already at lower temperatures. On the other hand, the uptake of excess oxygen due to doping represents a competing effect, since the tetragonal phase involves many repulsive nearest-neighbour oxygen pairs if the oxygen content in the Cu(1) plane is high. In order to study the orthorhombic-to-tetragonal phase transition in an open system order-disorder model[8,11] one may proceed as follows. The Free Energy $F(\{n_1\}, n_n, c, P_O, T)$ is minimized with respect to the order parameters (η_i), which refer to long-range and short-range order of the oxygen atoms:

$$\frac{\partial F}{\partial \eta_i} = 0, \quad (4)$$

with $\mu_{gas} = \mu_{sol}$ for the chemical potential of the oxygen. Solving these equations we obtain the η_i and the oxygen content in the Cu(1) plane $c = (1+y)/2$ ($-1 < y < 1$) as a function of P_O and T. (Impurities affect the superstructures in the Cu(1) plane occuring when oxygen is removed from the system).

However, the formation of small orthorhombic domains in the Cu(1) plane due to doping cannot be described by using only a few order parameters. Thus, to be able to incorporate the various oxygen redistributions around the impurities we use the Oguchi method[12]: We treat the oxygen-oxygen interaction energies ε_{00}, ε'_{00}, ε_{00}^{Cu} and ε_{00}^{M} (Fig. 3) within a cluster of oxygen atoms exactly, and the cluster is coupled to the remainder of the Cu(1) plane by an effective (molecular) field. Note, we assume a strong attraction ε_{00}^{M} linking two oxygen atoms through an impurity. Thus, in the ground state each impurity is surrounded by four oxygen atoms within the Cu(1) plane - in agreement with our results at T=0.

We then start from the partition function

$$Z = \sum_{i=1}^{n} \sum_{t_i=0}^{1} \exp\{-\beta H(t_1,\ldots t_n,c_1\ldots,\tilde{c}_n,\mu)\}, \qquad (5)$$

where the cluster Hamiltonian is given by

$$H = \sum_{\langle ij\rangle}^{cl} J_{ij}t_it_j + \sum_{\langle ik\rangle}^{--} J_{ik}(t_i - c_i)c_k + \mu \sum_{i=1}^{n} t_i. \qquad (6)$$

Here, $\sum_{i=1}^{n}$ runs over the oxygen sites within the cluster, $\sum_{\langle ij\rangle}^{cl}$ runs over all nearest- and next-nearest-neighbour pair sites in the cluster and $\sum_{\langle ik\rangle}$ runs over those pair sites, where one site belongs to the cluster

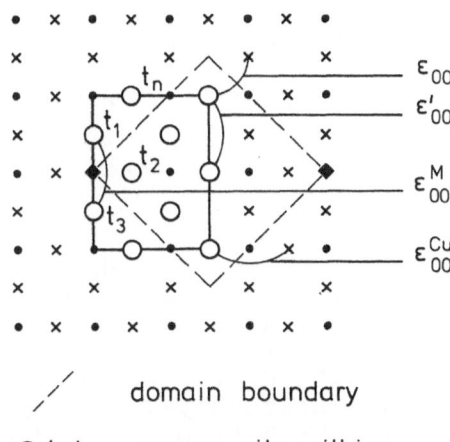

/ /   domain boundary

O (×)   oxygen site within (outside) the cluster

Fig. 3. A cluster of inequivalent oxygen sites around an impurity in the case of an uniform distribution of the impurities on the Cu(1) sites.

whereas the other does not. The J_{ij} are the interaction parameters, and the c_k are defined by $c_k = \langle t_k \rangle$ and denote the average occupation of an oxygen site. Note, the inclusion of the term $\mu(T,P_0)\sum_{i=1}^{n}t_i$ enables us to treat the system as an open system.

From this model we obtain the self-consistent equations for the average occupation numbers of various inequivalent oxygen sites $c_k (k = 1,\ldots,n)$,

$$c_k = \frac{1}{Z} \sum_{i=1}^{n} \sum_{t_i=0}^{1} t_k \exp\{-\beta H\}. \qquad (7)$$

Solving the coupled equations for the c_k we find how the occupation of the various oxygen sites changes with temperature. Thus, we can predict, how the entropy-driven transition to the tetragonal phase is affected by the presence of impurities in the Cu(1) plane.

SUMMARY

We have shown that impurities may induce strong oxygen redistribution in $YBa_2(Cu_{1-x}M_x)_3O_{7+y}$ systems. In particular, impurities cause domain formation, involving uptake of excess oxygen, which leads to a=b for the average lattice constants. Thus, superconductivity in alloys with a=b refers to systems consisting of a mixture of differently oriented orthorhombic domains. Presently, we are extending our model by performing electronic calculations of the ground state energies and by including the effects of finite temperatures.

We ackowledge helpful discussions with P. Jensen and M.E. Garcia.

REFERENCES

1. J.D. Jorgensen, M.A. Beno, D.G. Hinks, L. Soderholm, K.J. Volin, R.L. Hitterman, J.D. Grace, I.K. Schuller, C.U. Segre, K. Zhang, and M.S. Kleefisch, Phys.Rev. B 36, 3608 (1987).
2. P. Monod, M. Ribault, F.D. Yvoire, J. Jegoudez, G. Collin, and A. Revcolevschi, J.Physique 48, 1369 (1987); R.J. Cava, B. Battlog, C.H. Chen, E.A. Rietman, S.M. Zahurak, and W. Werder, Phys.Rev. B 36, 5719 (1987).
3. J.M. Tarascon, P. Barboux, P.F. Miceli, L.H. Greene, G.W. Hull, M. Eibschutz, and S.A. Sunshine, Phys.Rev. B 37, 7458 (1988).

4. Z. Hiroi, M. Takano, Y. Takeda, R. Kanno, and Y. Bando, Jpn.J.Appl.Phys. $\underline{27}$, L 580 (1988); G. Roth, G. Heger, B. Renker, J. Pannetier, V. Caignaert, M. Hervieu, and B. Raveau, Z.Phys. B $\underline{71}$, 43 (1988).

5. B.D. Dunlap, J.D. Jorgensen, C. Sergre, A.E. Dwight, J.L. Matykiewicz, H. Lee, W. Peng, and C.W. Kimball, Physica C, in press.

6. J. Ziolkowski, J. Solid State Chem. $\underline{57}$, 269 (1985); I.D. Brown, in Structure and Bonding in Crystals, Vol. 2, edited by M. O'Keffe and A. Navrotsky Academic Press, New York (1981).

7. p is given by p ≡ 3rx, where x is the impurity content in $YBa_2(Cu_{1-x}M_x)_3O_{7+y}$ and r is the number of impurities which occupy Cu(1) sites devided by the total number of impurities.

8. G. Baumgärtel and K.H. Bennemann, submitted to Phys.Rev. B.

9. H. Bakker, D.O. Welch, and O.W. Lazareth, Jr., Solid State Commun. $\underline{64}$, 237 (1987); Jeffry L. Tallon, Phys.Rev. B $\underline{39}$, 2794 (1989).

10. J.M. Sanchez, F. Mejia-Lira, and J.L. Moran-Lopez, Phys.Rev. B $\underline{37}$, 3678 (1988); L.T. Wille, A. Berera, and D. de Fontaine, Phys.Rev.Lett. $\underline{60}$, 1065 (1988).

11. H. Shaked, J.D. Jorgensen, J. Faber, Jr., D.G. Hinks, and B. Dabrowski, Phys.Rev. B $\underline{39}$, 7363 (1989).

12. J. Samuel Smart, Effective Field Theories of Magnetism, W.B. Saunders Company (1966), chapter 4.

CHARACTERIZATION OF TWINS AND TWIN DOMAINS OF YBaCuO

SINGLE CRYSTALS BY MICRO-RAMAN SCATTERING

P. Knoll[*]

Max-Planck-Institut für Festkörperforschung
Heisenbergstraße 1, D-7000 Stuttgart 80, FRG
and
Institut für Experimentalphysik der Univ.Graz
Universitätsplatz 5, A-8010 Graz, Austria

__Abstract.__ Several single crystals of $YBa_2Cu_3O_{7-x}$ with different twin structures are investigated by micro-Raman scattering. The comparison of the phonon line positions from different samples suggests a change in atomic positions due to the oxygen arrangement. Especially, a superconducting, nearly tetragonal crystal was found with a low lying O(4) vibration. The results are discussed together with Raman spectra of Al-doped crystals and lattice-dynamical calculations. In conclusion a model of the behavior of the O(4) vibration is suggested.

INTRODUCTION

Oxygen vacancies and distortions of the oxygen octahedra can reasonably alter the physical properties of perovskites. Therefore, it is not surprising that also in the new high-T_c materials the superconducting properties strongly depend on the oxygen content and some influence of the oxygen and oxygen-vacancy ordering is supposed. Even in the $YBa_2Cu_3O_{7-x}$ compound oxygen controls the physical behaviour in a wide range, from an antiferromagnetic insulator to a superconducting metal. Beside a number of standard techniques like neutron diffraction, electron microscopy, thermogravimetry, etc. which, more or less, directly investigate the oxygen content or arrangement, spectroscopic techniques like Raman scattering

---

[*]Alexander von Humboldt foundation fellow

have been established. As an example, the intensities and
positions of the observed Raman lines of $YBa_2Cu_3O_{7-x}$ have
been proven to be very sensitive to the oxygen content[1].

In a simple interpretation Raman spectra can be
taken as a finger-print method for a chemical analysis
and impurity phases have been found and identified in the
high T_c materials by this method[2]. Taking into account
the detailed physical mechanism of Raman scattering on
phonons one can obtain information about the phonon
behavior at the center of the Brillouin zone[3], in
resonant Raman experiments about the electronic structure
and by interpreting the intensity of the Raman lines
about the electron-phonon coupling.

Raman scattering also can be used, in principle, to
investigate the crystal structure and small distortions
of the atomic arrangement as it will be done in this
paper by two ways:

a) The polarization selection rules of incident and
scattered light strongly reflect the symmetry of the
atomic arrangement. Together with group-theoretical
treatments determining the Raman intensities for several
polarization directions yields information on the crystal
symmetry. Even questions about disorder, tetragonal to
orthorhombic phase transitions etc. can be clarified from
such experiments.

b) Small distortions of the atomic positions are seen as
a shift of the Raman lines which may be more sensitive
than X-ray, neutron or electron diffraction. By
comparison with lattice-dynamical calculations not only
changes in atomic positions but also changes of the
atomic environment can be identified.

An advantage of Raman scattering is the possibility
to investigate very small particles, crystals or parts of
a crystal, in principle, down to a size of the wavelength
of the laser used. Such a micro-Raman technique was
applied for most of the experiments reported in this
work. A 514.5nm laser beam was focussed through an
optical microscope to a spot of around 1μm diameter. The
scattered light was coupled through the microscope to an
OMARS 89 Raman spectrometer with an intensified diode-
array detector. Polarizers and polarized white light were
used in order to characterize the sample material by
means of polarization microscopy and to perform polarized
Raman experiments on distinct parts of the sample.
Additionally, conventional Raman measurements have been
performed on a Spex triplemate spectrometer with a
mepsicron detector (ITT, Fort Wayne, IN ,USA) also using
the 514.5nm Argon radiation.

COMPARISON OF DIFFERENT SINGLE CRYSTALS

The Raman properties of $YBa_2Cu_3O_{7-x}$ have been studied by several groups[3] and the five A_g modes are found to be at $120cm^{-1}$, $150cm^{-1}$, $335cm^{-1}$, $430cm^{-1}$ and $500cm^{-1}$. Lattice-dynamical calculations[4] have shown that these modes belong to the Ba, Cu(2), out-of-phase O(2) and O(3), in-phase O(2) and O(3), and O(4) vibrations along c, respectively. Thereby an orthorhombic structure (D_{2h}) is assumed and the atoms within the elementary cell are labeled as given in ref.2. The exact positions of these five modes vary from sample to sample and even the change in oxygen content has been proven to cause a shift of the lattice vibration frequency[1]. In particular, this is true for the $500cm^{-1}$ O(4) vibration which shifts from $505cm^{-1}$ (O_7) to below $480cm^{-1}$ (O_6). Although this shift is not understood in detail, it is sometimes used to determine the oxygen content. In the same way also the lattice constants (even the orthorhombic distortion) and the superconducting transition temperature strongly depend on the oxygen content. The question is whether the shift of the O(4) vibration and the change of the superconducting transition temperature are more a consequence of the oxygen content or of the orthorhombic distortion (which corresponds to the oxygen arrangement). Samples having Cu partly substituted by Fe, Al, Co, Ga etc.[5,6], become tetragonal and are still superconducting. These results sugest that the superconducting transition is dominated by the oxygen content and the orthorhombic distortion is of less influence. If the same is true for the shift of the O(4) vibration a strong correlation between superconducting transition temperature and O(4) phonon frequency should excist. Results on the substituted sintered ceramics[7] suggest that the O(4) vibration better correlates with the orthorhombic-tetragonal transition than to the superconductivity (or oxygen content). Also, in samples substituted with Ca a clear reduction of the orthorhombic distortion without any dramatic change of the superconducting transition temperature was observed[8] although the O(4) $500cm^{-1}$ Raman mode shifts down to $480cm^{-1}$. These measurements suggest that the O(4) vibration is more related to the oxygen ordering (which also determines the orthorhombic distortion) than to the oxygen content. In order to prove this without impurities as Ga, Fe etc. the position of the $500cm^{-1}$ Raman mode has been compared to the superconducting transition temperature for several single crystals where different oxygen arrangements are assumed. Polarization microscopic photographs of such crystals are shown in Fig.1 demonstrating the influence of oxygen

ordering in the different twinning structures due to
different preparation conditions. Usually the size of the
twins (around some 100nm) is too small to be seen in the
light microscope (Fig.1a) but as there are two equivalent
twin boundary directions, 110 and 1$\bar{1}$0, twin domains are
formed making the contrast in Fig.1a. The size of the
twinfree regions can come up to some μm (Fig.1b) and also
some crystals are found looking tetragonal. Raman spectra
exhibiting the O(4) 500cm^{-1} vibration of such crystals
together with a single crystal found in sinter ceramics,
Al doped single crystals, and a sputtered film are shown
in Fig.2 in comparison with the superconducting transition temperatures as obtained from electrical conductivity and magnetic measurements. No correlation of T_c and
phonon frequency was found. More it seems that also at
the single crystals the shift of the 500cm^{-1} phonon is a
consequence of the loss of orthorhombic distortion.

The nearly tetragonal crystal is quite interesting.
Laue diagrams[9] and X-ray measurements on a four angle
diffractometer[10] show that the whole crystal is in agreement with a tetragonal elementary cell with a difference

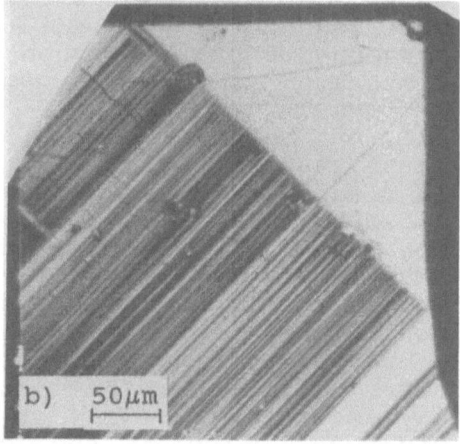

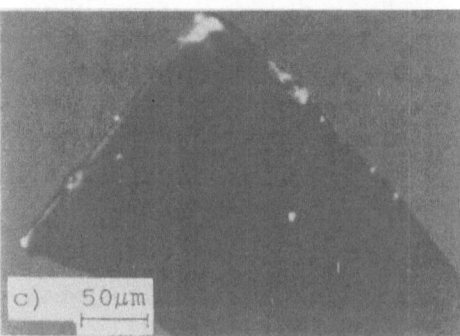

Fig. 1. Twinning structures
of different single
crystals. a) Submicron
twinning with twindomains
in (110) and (110) (grown
at TU Graz) b) large
twinfree regions with
twinboundaries in one
direction (grown at Argonne
Nat.Lab.) c) nearly
tetragonal crystal (grown
at TU Graz)

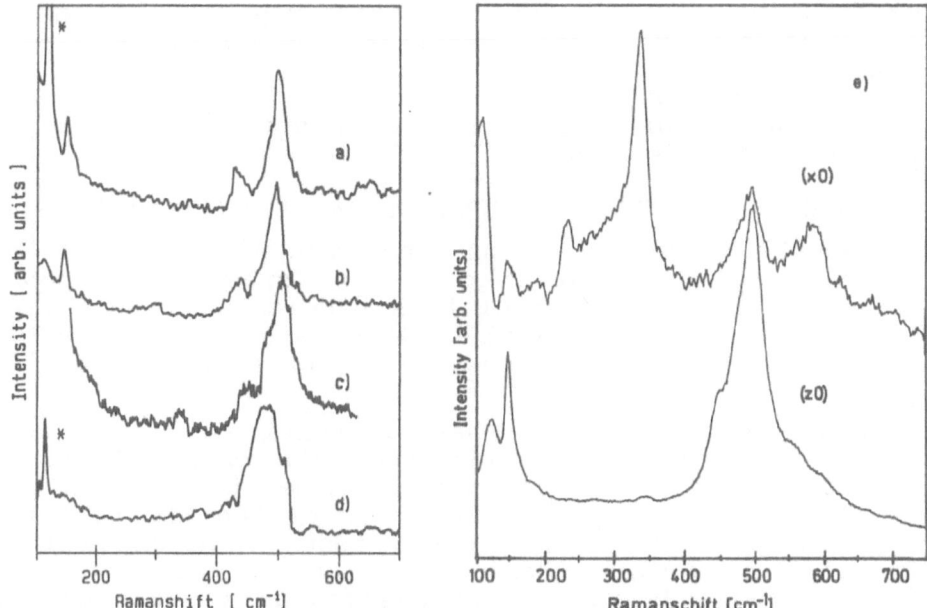

Fig. 2. Comparison of the O(4) vibration for different samples a) single crystal of sinter ceramics (O(4) at 502cm^{-1}, T_C=90K) b) single crystal from Fig.1b (O(4) at 500cm^{-1}, T_C=92K) c) sputtered film (O(4) at 500 cm^{-1}, T_C=70K) d) single crystal from Fig.1c (O(4) at 480 cm^{-1}, T_C=85K) e) single crystals of YBa$_2$Cu$_{2.9}$Al$_{0.1}$O$_{7-x}$ (O(4) at 495cm^{-1}, $T_C \sim$60K)   *indicates laser plasma lines

between the a and b axis of less than 0.02Å. The c axis did not show a significant change compared to the O$_7$ material. Even the shift of the O(4) vibration demonstrates that there must be a truly microscopic change of the structure and that the tetragonal symmetry is not simply an average over orthorhombic parts too small to be detected by X-ray. A detailed micro-Raman study on a freshly broken edge revealed that the position of the Raman line is almost constant over the thickness of the sample. The composition of the crystal was proven by EDAX and no indication of impurities was found except for a very low amount of Al. As only exists experimental Raman data of Al contaminated 123 ceramics[11] measurements on single crystals of the composition YBa$_2$Cu$_{2.9}$Al$_{0.1}$O$_{7-x}$[12] were performed and are shown in Fig.2e for two different incident polarization directions. The onset of superconductivity of that crystal is 60K and the crystal structure is nearly tetragonal. The shift of the O(4) vibration is small compared to the large shift down to $480 cm^{-1}$ of the previously reported crystal which surely has a smaller amount of Al. It can be concluded that Al by itself does not explain the observed large frequency shift. Changes in atomic positions due to the oxygen arrangement are assumed.

INFLUENCE OF OXYGEN ARRANGEMENT ON RAMAN PROPERTIES

As the arrangement of the oxygen seems to be critical for the observed Raman spectra a short theoretical investigation of this problem will be given. Results of group-theoretical treatments are summarized in table 1 and the several cases of oxygen arrangement will be discussed separately.

a) Orthorhombic-Tetragonal Phase Transition

Removing the O(1) oxygen from the chains triggers the orthorhombic to tetragonal phase transition. An equivalent probability of occupation of the two possible oxygen sites in the basal plane results in an average tetragonal structure. Besides a random occupation of these two sites (which can be assumed as a state not in thermal equilibrium) also the stability of ordered tetragonal O_7 phases was shown theoretically[13]. It requires, however, effective O-O-pair interactions in a range which do not make too much physical sense without additional assumptions such as defects, impurities etc.. The last two rows of table 1 show the change of irreducible representations (and as a consequence the change of the Raman polarization selection rules) for the tetragonal to orthorhombic transition without oxygen in the basal plane. For Raman spectroscopy the interesting feature is the change from an A_g (D_{2h}) phonon to a B_{1g} (D_{4h}) one. But the orthorhombic distortion is so small that this B_{1g} type is also observed in the orthorhombic structure. Therefore, adding oxygen in the basal plane does not significantly influence the Raman selection rules. In exact tetragonal symmetry the next possible oxygen content is O_8 and will add $1A_{2u}+1B_{2u}$(silent)$+2E_u$ modes, all not observable in Raman scattering.

Recently a lattice-dynamical shell model has been successfully applied to the O_7[14] as well as to the O_6[15] material. Within this model reducing the orthorhombic distortion of the O_7 material by equalizing the lattice parameters a and b has little influence on the Raman frequencies[16]. Only an increase of the B_{1g} like $335 cm^{-1}$ frequency (0.6%) and nearly no change of the O(4) vibration (0.05%) are observed. Therefore, the orthorhombic to tetragonal lattice distortion itself cannot be responsible for the observed large shift of the

$500cm^{-1}$ mode. In the same way only randomly occupying
both possible oxygen sites on the basal plane ($O(1)$ and
$O(5)$) by the same amount will also not explain the
observed shift as no interatomic distances change. The
only possibilities are the Born-Mayer constants, which
may slightly change because of the loss of square-planar
arrangement of the oxygen, and different z positions of
$O(4)$ and Ba due to the different arrangement of the
oxygen. Of course, the $O(4)$ vibration is very sensitive
to the $O(4)$ position. Calculations with slightly changed
Born-Mayer potentials of the Ba have shown that also Ba
has some influence on the z motion of $O(4)$. Changing the
corresponding interatomic distance by 1% will shift the
frequency of the $O(4)$ vibration by 0.2%.

b) Disordered Oxygen Vacancies
        Introducing oxygen vacancies along the $Cu(1)-O(1)$
chains will reduce the symmetry dramatically. The
translational symmetry breaks down and also all mirror
and rotational axes perpendicular to y are lost.
Calculations of the $O(1)-Cu(1)$ vibrations show only a
weak dispersion along k_x and k_z[14] which indicates that
there is only a weak interaction between the chains.
Therefore, the general case where the vacancies are
randomly distributed over all chains can be approximated
by vacancies only randomly disordered along the chain and
all chains have the identical vacancy distribution. This
will recover the translational symmetry in x and z
direction and only reduces the point-symmetry to C_{2v} with
the C_2 axis in y direction. The change in irreducible
representations is illustrated in table 1 (rows 1 and 2).
As expected, also the original u modes become observable
in Raman scattering because of the loss of inversion
symmetry. Most interesting are the new A_1 modes as they
are usually stronger in intensity than the non-totally-
symmetric ones. They originate exclusively from the B_{2u}
infrared-active modes.
        There are some uncertainties in the interpretation
of the Raman spectra, so far. The modes at $580cm^{-1}$ and
$220 cm^{-1}$ have been observed with different strengths on
several samples. It was demonstrated that laser
irradiation[17] and increasing the temperature[18] can
enhance these modes. Infrared modes activated by symmetry
breaking due to oxygen vacancies has been suggested as an
explanation. The polarization selection rules clearly
identify these modes as totally symmetric. Unfortunately,
no consistently measured frequencies of the B_{2u} and B_{3u}
modes exist. Only calculated data are available but, as
these calculations are in good agreement with all other
experimental observed frequencies in the O_7 as well as in

Table 1. Factor group analysis and group correlations for different oxygen content and ordering for $YBa_2Cu_3O_y$

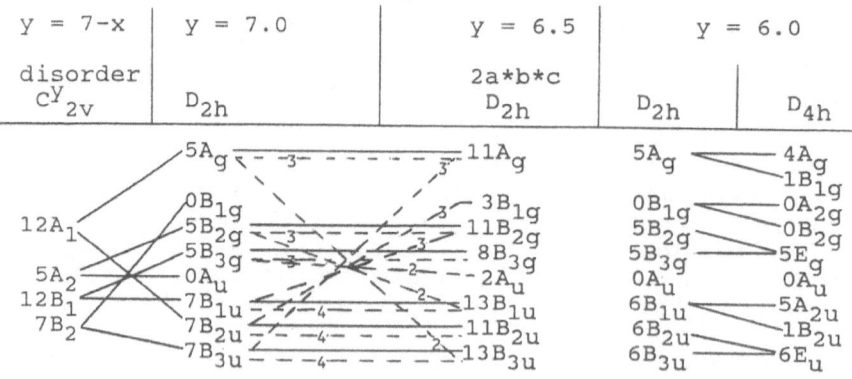

the O_6 material[19], they seem to be reliable. The only possible B_{2u} mode which can cause the 580cm^{-1} "disorder" mode is the $O(1)$ vibration in y direction which is calculated[14] at 573cm^{-1}. Oxygen dependent IR measurements have also seen structure in the range of 530 - 580cm^{-1} which was assigned to the $O(1)$ vibration in good agreement with theoretical results[20]. Also Thomsen and Cardona[3] suggested that this mode may explain the Raman feature at 580 cm^{-1}. The authors suggest the corresponding $O(1)-Cu(1)$ in-phase vibration in y direction to be responsible for the 220cm^{-1} Raman peak in agreement with the considerations given here. As the frequency of this motion (calculated at 103cm^{-1})[14] is quite far away another possibility will be suggested. By some amount also the second oxygen site ($O(5)$) in the basal plane can be occupied. The vibration of this $O(5)$ oxygen in y direction is a Cu-O bending (calculated at 168cm^{-1}) and has B_{2u} symmetry. This will explain the occurence of the 220cm^{-1} feature in a more consistent way and is favored by this paper.

c) Ordered Oxygen Vacancies and Superstructure

It has been reported that for $O_{6.5}$ a highly ordered superstructure occurs which yields a doubling of the elementary cell in the x direction[21]. The influence on the Raman polarization selection rules has been investigated group theoretically and is listed in table 1 (rows 2 and 3). All q=0 vibrations of the small elementary cell have their corresponding partners in the doubled cell. The additional q=0 motions of the doubled cell arise from the folding of the edge of the Brillouin zone (X point) to the center (Γ point). This is indicated

by dashed lines and the numbers indicate how many zone-
edge phonons of a branch with specific symmetry at the
zone center contribute to the q=0 vibrations of the
doubled cell. (No mixing of normal coordinates is
assumed.) Adding one oxygen to the doubled elementary
cell increases the oxygen content from $O_{6.5}$ to O_7 and
$1B_{1u}+1B_{2u}+1B_{3u}$ has to be added to the irreducible
representations. Of most interest are the 11 A_g modes
which consist of the 5 A_g modes of the simple cell and 6
folded zone edge vibrations, 3 from A_g branches and 3
from B_{3u} ones. From this considerations one would expect
that the vacancy ordered case should be well
distinguishable from the disordered case. First Raman
measurements on such highly vacancy ordered, but
unoriented samples have been reported by Blumberg et
al.[22]. Indeed new Raman bands occur but also bands at
$229cm^{-1}$ and $591cm^{-1}$ very similar to the disordered case.
Unfortunately, the Raman features in both cases are to
broad to distinguish clearly between ordered and disorded
case from these measurements.

CONCLUSIONS

     In conclusion, following model of the O(4) vibration
can be given:

a) The Oxygen Deficient Case
     Removing the oxygen of the basal plane also reduces
the repulsive potential for the O(4) due to the absence
of the repulsive O-O Born-Mayer potentials and the
repulsive O-O Coulomb interaction. This is thought to be
the main reason for the frequency decrease.

b) Substituted Material
     Substituting for Cu(1) trivalent metal atoms[5] will
cause the oxygen to be more concentrated around this atom
(mainly in octahedral structure). Although in total more
oxygen can be inside the sample, around the Cu(1) there
is less oxygen which, additionally, is tetragonally
arranged. The doping of the CuO_2 planes is not effected
too much by this different arrangement of oxygen. The
reasons for the lowering of the O(4) vibration can be the
reduced oxygen content around the Cu(1) atoms as
discussed in paragraph a) and the tetragonal arrangement
of oxygen as described in paragraph c).

c) Nearly Tetragonal Single Crystals
     The tetragonal arrangement of the oxygen is thought
to be responsible for the higher symmetry and the
lowering of the O(4) vibration. The superconducting
properties will still remain. The shift of the O(4)

vibration can either be due to destroying the square
planar oxygen environment of the Cu(1) or to a shift of
atomic positions O(4) and Ba along z as a reaction to the
new oxygen distribution in the basal plane.

ACKNOWLEDGMENT

I thank H.W.Weber, J.Z.Liu from Argonne Nat. Lab.,
G.Leising and B.Gegenheimer for sample material and
W.Kress, C.Thomsen and M.Cardona for valuable
discussions. M.Pressl and B.Stadlober is greatly acknowledged for excellent experimental help. This work was
supported by the Alexander von Humboldt foundation and
the Fonds zur Förderung der Wissenschaften in Austria.

REFERENCES

1. see e.g. R.M.Macfarlane et al. Phys.Rev.B38,284(1988)
2. P.Knoll and W.Kiefer, in "High Tc Superconductors"
   ed.H.W.Weber, Plenum press, p.121(1988)
3. for a recent review see C.Thomsen and M.Cardona, in
   "Physical properties of High Temperature
   Superconductors" Ed. D.M.Ginsberg,
   World Scientific,Singapore, p.409(1989)
4. R.Liu et al., Phys.Rev.B37,7971(1988)
5. J.M.Tarascon et al. Phys.Rev.B37,7458(1988)
6. G.Xiao et al.Phys.Rev.Lett.60, 1446(1988)
7. H.Kuzmany, E.Faulques, M.Matus, S.Pekker in "Studies
   of High Temperature Superconductors", Ed.H.Narlicar,
   Nova Science Publishers, Vol.3, p.299(1989)
8. G.Leising et al. Physica C 153, 886(1988)
9. O.Leitner and G.Leising, unpublished
10.F.Belaj, unpublished
11.P.B.Kirby et al. Phys.Rev.B36, 8315(1987)
12.B.Gegenheimer, G.Jasiolek, A.Pajaczkowska,(submitted)
13.L.T.Wille, D.deFontaine, Phys.Rev.B37, 2227(1988)
14.W.Kress et al. Phys.Rev.B38, 2906(1988)
15.C.Thomsen et al. Sol.State Commun.65, 1139(1988)
16.W.Kress, unpublished
17.P.Knoll and W.Kiefer, Proceedings of the XIth
   International Conference on Ramanspectroscopy,
   London, J.Wiley, p.415(1988)
   D.Mihailovic and J.Solmajer, Phys.Rev.B (submitted)
18.K.F.McCarty et al. Phys.Rev.B38, 2914(1988)
19.L.Genzel et al. Phys.Rev.B (submitted)
20.H.Kuzmany et al. Sol.State Commun.65, 1343(1988)
21 T.Takabatake et al. Physica C, 152, 424(1988)
22.G.E.Blumberg et al. Sol.State Commun.70, 647(1989)

OXYGEN DISORDER EFFECTS IN HIGH T_c SUPERCONDUCTORS, 1989

K.H. Bennemann

Institute of Theoretical Physics
Freie Universität Berlin
Arnimallee 14, 1 Berlin 33, W. Germany

When I was asked to prepare a summary I tried to recommend someone else to summarize this fine conference. I used to argument that I didn't understand high T_c-superconductivity. However, this was not accepted as an exclusive argument.

When I continued to argue that this would not permit me to behave like an oxygen deserting his O(1) or O(4) site ocasionally, by which I was referring to relaxing walks in the nearby beautiful park, I was told to attend first the lectures and then to take a refreshing walk.

Similarly, it may be reasonable to discuss first the many good lectures by forgetting the main motivation for all the studies, namely high T_c superconductivity. Then, it seems fair to me to say that many interesting and good talks were presented at this conference.

As correctly suggested by the title of the conference the

behaviour of oxygen

is of vital importance for the properties of High T_c-materials. The various studies may be characterized as illustrated by the figure.

One may say, there is <u>impressive progress</u>, both experimental and theoretical, with respect to our understanding of the <u>lattice</u> and <u>electronic structure</u> of the complex high T_c-material, quality of the samples (sample preparation, conditions) and the dependence of the <u>various lattice phases</u> on <u>oxygen distribution</u>. There is

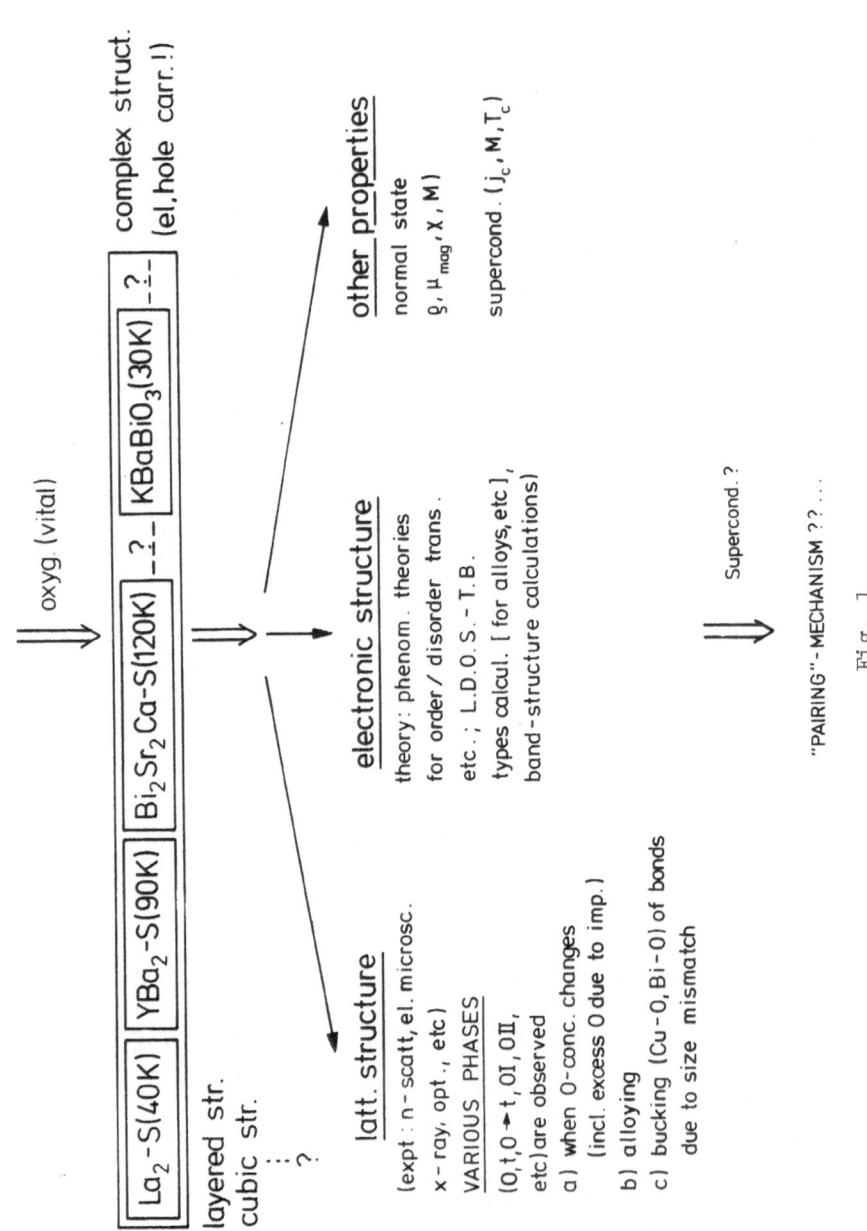

Fig. 1

probably no doubt that this is useful for reaching the ultimate
goal of understanding the physical mechanism for high T_c superconductivity.

However, one must admit that presently there is no real clue
linking
                    structure $\leftrightarrow$ high T_c superconductivity.

The Cooper-pairing (?) mechanism remains unclear, even so one may
suspect that definite results on:

T_c (n_h) or T_c ($N(0)$, $<\omega_{ph}>$,..),
T_c (δ), (s. plateau in $YBa_2..O_{7-\delta}$)
T_c (Z_n, Ga, ..)
T_c ($Ba_{1-x}M_xBiO_3$)
Isotope effect, etc.,

contain important hints already (as well as other experimental
results).

It is important fo find out some of the factors which explain
why T_c varies so much,
for La_2-S $\rightarrow$ YBa_2-S $\rightarrow$ Bi_2Sr_2-S $\rightarrow$?

Here, calculations of electronic structure (s. Mattheiss) and
systematic comparison with related system exhibiting low T_c-super-
conductivity seems very important, as well as studies of $N(0,\delta)$,
n_h, etc. (s. Lambin, Anda-Morán-López,...).

One should mention, that during this conference several results were presented suggesting (Sinha, Chien, Barboux,..) that
high T_c superconductivity and magnetism do not like each other.

What is the role of the chain (in 123's) for superconductivity (?), do Bi-O layer contribute to superconductivity, why
T_c α n_h (sometimes), oxygen displacements, bond buckling, (Egami!)
why?

Is there besides superconductivity and magnetic activity
($\mu(\delta)$?,..'),

                    piezoelectric behaviour?

and a sensitive
                    electronic polarizability?

(of Cu-O bonds) and what is the dependence on oxygen distribution?

These and many other questions might help to motivate new
(conferences!) studies and more refined studies concerning the

atomic <u>displacements</u> and <u>environment</u> of <u>superconductive-active</u>
<u>sites</u> (s. Egami, Chien [induced magnetic moments by impurities],
et al.), their valency, special local phonon modes, etc.

CONFERENCE HIGHLIGHTS (personal view):

   The results on the <u>lattice structure, various phases</u> (o, t,
superstructure), controlled sample preparation were very impressive (s. talks by Jorgensen, Sinha, Alario-Franco,..., and in
particular by Amelinckx).

   Similarly, the <u>band-structure calculations</u>, and here in particular the discussions by Mattheiss (somewhat comparabele to the
impressive experimental results presented by Amelinckx) were also
very impressive. Here, one should also mention the tight-binding
type calculations by Lambin, particularly suited for alloys, and
the calculations by Anda, Morán-López using the <u>Hubbard-Hamiltonian</u>
and thus including the possibly important <u>e-e correlation</u> effects
($N(0)$, h_h, etc.).

   The <u>"phenomenological"</u> theories using effective pair like
inter-atomic interactions ($v_{ij}(\delta,..)$) seem to be in accord with
experiment and rather successful in determining various lattice
phases. This was well discussed by de Fontaine. As he pointed out
theory may be <u>plagued</u> by <u>approximations</u> (molecular field approx.?)
similarly as experiments by sample preparation.

   Despite many success, there remain <u>problems, questions</u>:

(a) Mobility of oxygen (T>?), route for O-diffusion
(b) O-vacancy formation, interstitial sites
(c) Extended defects, domain formation
(d) Local O-rearrangement due to impurities
(e)

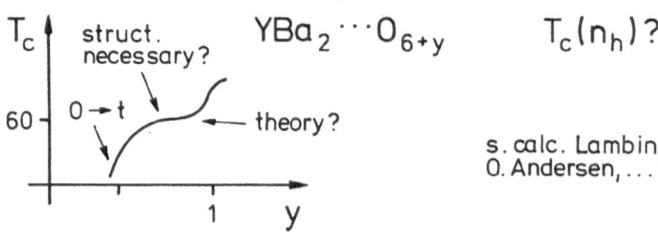

Fig. 2.

(f) Interfaces, thin films (O-concentration profile)(tunnel-func-
   tions)
(g) n_h on Cu, O..?
(h) Range and origin of super-structures.

   Phenomenological theories using effective atomic pair-inter-
actions (V_{ij}) for determining various lattice phases should be
extended to include lattice distortions and interactions. V_{ij}
should be derived from electronic theory (for example, using tight-
binding moment expansion, recursion techniques, and band-structure
calculations [Mattheiss, et al.] as a guide).

   This is necessary to link various properties with electronic
structure.

OTHER STUDIES

   Effects of substituting (in 123 systems, for example) for Cu:
Zn, Ga, or transition-metal atoms (s. Barboux,...) on T_c seem to
me very interesting and important. This I found in particular
convincingly demostrated by Chien's results. Such experiments
should be supplemented by local density of states type electronic
calculations.

   Magnetism is like superconductivity a many body effect (due to
e-e correlations). Therefore, studies of $\mu_{Cu}(\delta)$, $T_N(\delta)$-magnetic
interactions are very important ($T_N(n_h \rightarrow 0) \rightarrow ?$), s. talk by B. Maple,
in particular.

   One should forgive me for not mentioning all the other interesting results presented at this conference, but time limits me.

   In conclusion, let me thank, on behalf of all attendants, the
organizers of this conference for selecting so many interesting
and good talks. Certainly a prerequisite like for the good quality
of this conference, Oxygen is for high T_c-material. I wish all of
you good new ideas and future success.

INDEX